Rainbows
Daniel MacCannell

REAKTION BOOKS

To Emily Amelia Juliet

Published by Reaktion Books Ltd
Unit 32, Waterside
44–48 Wharf Road
London N1 7UX, UK
www.reaktionbooks.co.uk

First published 2018

Printed and bound in China by 1010 Printing International Ltd

A catalogue record for this book is available from the British Library

ISBN 978 1 78023 920 0

CONTENTS

1 What Rainbows Are and How They Work

Rainbows are not objects. They cannot be approached or touched, and cannot be said to have any actual size. Every rainbow is a distorted image of the Sun, and like the Sun, it would appear as a full circle if the surface of the Earth did not get in the way. Each rainbow is very much 'in the eye of the beholder': multiple observers can see similar rainbows in the same approximate location at the same time, but no two people can actually see the 'same' rainbow, even if they are standing directly beside each other. This is because the centre of each rainbow-circle is located on an imaginary line that could be drawn from the Sun into the back of the individual observer's head, out again through his or her eye, and to the shadow of the head on the ground. As such, the commonplace children's book and advertising image of a rainbow beside or in front of the Sun could never be seen in reality, except perhaps on a planet with more than one sun. Moreover, rainbows are created by the bending of sunlight through drops of water in the air, and such drops are always in motion – generally, towards the ground. Thus, even an unmoving observer is not seeing the 'same' rainbow from one instant to the next, but rather a lengthy series of very similar rainbows, created by a succession of new falling water particles as they pass through the portion of sky that is relevant to rainbow production from the standpoint of that individual person.[1] Nevertheless, for reasons that will be explained below, the bow always measures 2 degrees in thickness/height, and always appears at an angle of 40–42 degrees above the aforementioned imaginary line through the observer's head.

Rainbow and complex clouds after intense rain near Stockton, California.

7

Thousands of images like this one have been created, but in reality a rainbow and the Sun could never be seen at the same time.

It follows from all this that the rainbow appears largest when the Sun is lowest in the sky – that is, passing the horizon at sunrise or sunset. A person at sea level can only see the rainbow as a half-circle when the Sun is at the horizon, while to see it as a full circle requires him or her to be at a very high vantage point, typically on a mountaintop or in an aircraft. It also follows that the higher in the sky the Sun is at the moment of observation, the smaller the rainbow will appear; and when the Sun moves more than 42 degrees above the horizon, the rainbow must necessarily disappear altogether. Consciously or not, it may have been for this reason that the best-selling British science fiction author Douglas Adams (1952–2001) chose '42' as his comedic 'Answer to the Ultimate Question of Life, the Universe and Everything'.[2]

It became accepted in European scientific circles in the seventeenth century – though not before then – that the rainbow has seven colours, with the innermost band being violet, followed as one looks upwards by indigo, blue, green, yellow, orange, and finally red at the top. In fact, its colours shade into one another continuously, as black-and-white photographs of rainbows reveal. It is human perception – partly biological, and partly learned – that groups these fine colour gradations into seven (or some other number), probably via a cognitive process known as metamerism, in combination with cultural factors such as the number of different words for colours there are in the language that one speaks.[3] Metamerism refers to the fact that the normal human eye contains only three types of colour receptors, known as cone cells, with each type especially sensitive to a particular range of wavelengths of light; these sensitivities are strongly overlapping, but are centred around 440 nanometres (nm) (blue), 540 nm (green) and 570 nm (yellow/orange). It follows that it is more difficult for us to perceive red light (700 nm) than the other aforementioned colours; and where the wavelength of light is beyond 810 nm, we have no ability to perceive its colour at all. Certainly, due to this biological structure – some might say defect – of the human eye, we are able to see the 'same' colour as the product of quite different combinations of wavelengths of light, and are also tempted to see 'clumps' of colour where objective measurement finds none. Nevertheless, the precise reasons for our perception of the rainbow as a set of coloured bands remain the subject of speculation, as Ari Ben-Menahem explains:

However distinct they seem, the separateness of the coloured bands that make up the rainbow is produced in the mind.

Our knowledge of what goes on between the eye and the brain when one sees a rainbow is pretty much in a state of flux. Which part of what we see is due to physical factors, and which is due to purely entoptic [within-the-eye] reasons ... is still unknown.[4]

All true rainbows are produced by the interaction of sunlight with water droplets – often but not always raindrops – in the atmosphere. As well as rain, bows are frequently observed in mist, fog, dew and spray. Our understanding, via computer simulation, of precisely how the shapes and sizes of these water particles affect a bow's size, precise hues, visual sharpness and apparent luminosity is in its infancy. That being said, we do know that only relatively small raindrops contribute to the broadly

Diagram from *Gray's Anatomy* (1918) illustrating the relation of the rods and cones to other neurons within the human retina.

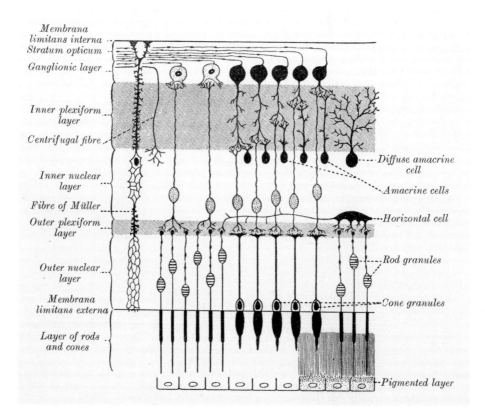

Membrana limitans interna
Stratum opticum
Ganglionic layer

Inner plexiform layer

Centrifugal fibre

Inner nuclear layer

Fibre of Müller
Outer plexiform layer

Outer nuclear layer

Membrana limitans externa

Layer of rods and cones

Diffuse amacrine cell

Amacrine cells

Horizontal cell

Rod granules

Cone granules

Pigmented layer

Fogbow at the Summit Research Station, Greenland.

horizontal 'crown' of the rainbow arch, whereas all sizes of drops contribute to its 'legs', and this is what makes the bow so often appear brighter nearest to the ground.[5] The size of a raindrop varies between 200 and 2,000 micrometres (μm), while visible fog droplets may be as tiny as 5 μm. It is this smallness of the droplets that renders so-called 'fogbows', in comparison to ordinary rainbows, 'broad, but . . . feeble' and with 'faint' colours that are generally perceived as white.[6] Rainbows produced by moonlight, known as 'moonbows', contain the same colours in the same order as regular rainbows, since moonlight is merely reflected sunlight. However, with each reflection, a considerable amount of light is 'lost', and this is especially so in the case of objects (including the Moon) that are not smooth. This renders moonbows so faint in comparison to normal rainbows that their colours can easily pass unnoticed. Ice crystals in the atmosphere can also produce a range of rainbow-like phenomena that are described briefly at the end of this chapter.

It took centuries of experimentation and debate to establish that the primary rainbow is produced by a combination of two refractions and one reflection at the level of the individual raindrop, with most early observers arguing for an all-reflection or all-refraction rainbow, and/or one that was produced not by drops but at the 'macro' level of whole clouds – or even the 'dome of heaven' itself. These debates are described in detail in Chapter Two; but the conclusions reached in the seventeenth century still broadly hold.

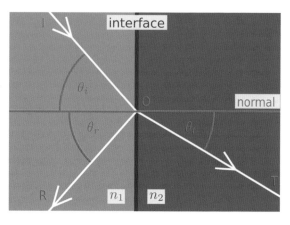

Diagram of reflection (I-R) and refraction (I-T), where n_1 and n_2 are substances with different refractive indices.

Crucially, a water drop is reflective on its inner as well as its outer surfaces. When sunlight strikes a raindrop, some of it is reflected from the outer surface, and thus (from the point of view of rainbow production) lost. But some of it enters the drop, at which point refraction – the optical process that can make one's legs appear bent or disconnected when entering a swimming pool – comes into play. The light that does enter the drop is first refracted – about 13 degrees downwards, in the case of rays entering at the 'mid-latitude' of a drop's 'northern hemisphere', and roughly 13 degrees upwards, in the case of rays hitting the same spot in its 'southern hemisphere'. The root cause of refraction is that water and air differ sharply in terms of density. This causes light to slow down dramatically when moving from air into water, from around 300 million metres per second to about 226 million metres per second, and thus to change direction slightly.[7] A decent analogy for this is the way in which a car would tend to swerve to the left if the wheels on its left-hand side suddenly encountered wet cement, and those on its right-hand side did not. Unlike angles of reflection, angles of refraction are not fixed.

They vary not only depending on whether the denser medium is water or glass, warm or cold water, and so forth, but – enormously – according to how obliquely the sunlight strikes the water's surface (the 'angle of incidence'): from less than 12 degrees if the ray strikes the water at a 15-degree angle, to more than 40 degrees in the case of a 60-degree angle of incidence. The table below sets out some representative values obtained from a single experiment:[8]

Angle of Incidence (degrees)	Angle of Refraction (degrees)
0.0	0.0
5.0	3.8
10.0	7.5
15.0	11.2
20.0	14.9
25.0	18.5
30.0	22.1
35.0	25.5
40.0	28.9
45.0	32.1
50.0	35.2
55.0	38.0
60.0	40.6
65.0	43.0
70.0	45.0
75.0	46.6
80.0	47.8
85.0	48.5

Having been thus bent on entering a raindrop, each ray of sunlight then encounters the drop's 'back wall', where it is reflected
at the same angle as it struck it, that is, to 29 degrees downwards
if it came from 29 degrees above, and so forth. The ray is then
refracted a second time, when and because it moves out of the
raindrop and back into the less-dense air.

The upshot of all this bouncing and bending is that there
is a particular point in what might be called the high northern
latitudes of such a water drop, above which all rays of sunlight
entering the drop eventually become concentrated at roughly
the same angle as one another when they exit the drop's lower
hemisphere. Now, heading back roughly in the direction of the
observer and the Sun, the rays have ceased to be exactly parallel, but remain in a tight, *almost* parallel formation, and always
pointing around 41 degrees below the line along which they first
entered the drop.

We say 'around' 41 degrees because there are also, crucially,
small colour-dependent differences in how much refraction
occurs, a phenomenon known as dispersion. Dispersion causes

Rainbow and secondary rainbow over Banchory, Kincardineshire, 2014.

blue-violet light to emerge from the drop at around 40 degrees, red light at about 42 degrees and all the other rainbow colours at points between. This explains the primary bow's 2-degree thickness or height, as well as the fact that it is multicoloured rather than white.[9]

A rainbow is often encircled by a larger but weaker 'secondary' rainbow that has its colours reversed – that is, red innermost and violet outermost – and this secondary bow occupies the space between 51 and 54 degrees above the Sun/observer line (that is, is 1 degree or about 50 per cent thicker than the primary bow). It is caused by light rays that have been reflected twice by the inner surfaces of the raindrop rather than once, but which still undergo just two refractions – one on entering and the other on leaving. Tertiary, quaternary and quinary rainbows also exist: the tertiary and quaternary appear behind the observer's back, and the quinary appears between the primary and secondary bows. However, as each such iteration requires

Full-arc double rainbow over the Stratford Canal Basin, 2013.

one additional within-drop reflection, these bows are almost inevitably too faint to be seen. Though up to two-hundredth-order rainbows have been produced under laboratory conditions using laser beams, no naturally occurring tertiary or quaternary rainbow was successfully photographed until 2011;[10] and the first confirmed photo of a natural quinary rainbow appears to have been taken just two years earlier.[11]

Tertiary, quaternary and quinary bows are not to be confused with *supernumerary* bows, extra stripes (usually of purple or green) that are sometimes seen immediately beneath the primary bow, as in our photograph overleaf taken near Aviemore, Scotland. These cannot be accounted for by additional reflections, or indeed any other feature of geometric optics, which have proved 'totally unable to explain them'.[12] Rather, they result from destructive interference, the process by which two waves or pulses – whether of light, sound, water or what have you – momentarily cancel each other out when moving in opposite directions. Indeed, the existence of supernumerary rainbows was crucial to the formulation of the wave theory of light by Thomas Young (1773–1829) in the first decade of the nineteenth century.[13] Modern simulations suggest that supernumeraries are most likely to be seen when atmospheric water drops are quite

James Gillray caricature of Thomas Young's experiments, 1802.

An all-red rainbow can sometimes be seen when the Sun is extremely low in the sky. This is due to the longer passage of its rays through the atmosphere, where short-wavelength greens and blues are scattered more strongly by dust and other atmospheric particles than longer-wavelength reds.

small and uniform in size – ideally, between 650 and 750 μm in diameter.[14]

Likewise, only the wave theory of light can explain why all-red rainbows are sometimes seen at sunrise and sunset. This is because the strongly slanting angle of sunlight at these times of day causes it to travel an unusually long way within the Earth's lower atmosphere, where its waves

are scattered by air molecules and dust. Short wavelength blues and greens are scattered most strongly leaving the remaining transmitted light proportionately richer in reds and yellows.[15]

Like a rainbow or moonbow, a 'glory' or 'pilot's glory' is caused by atmospheric water droplets and is centred on the shadow of the observer's head. However, it is much smaller than a rainbow, at between 5 and 20 degrees, and generally takes the form of a series of concentric rainbow-coloured circles, each with red outermost. Most importantly, it can only be seen if the water particles in question are at a lower altitude than the observer. Its cause is not definitively known, though the Brazilian physicist Herch Moysés Nussenzveig proposed an explanation based on the phenomenon of quantum tunnelling, and partly for that

Rainbow with
supernumeraries
over Aviemore,
Inverness-shire, 2007.

contribution received the American Optical Society's Max Born Award in 1986.

Most other rainbow-like phenomena that occur in the sky are caused by ice crystals rather than water drops, and are thus considered quite distinct from rainbows, scientifically speaking. Ice in clouds at higher altitudes is commonplace, falling as rain if it melts, and as snow if it does not. For reasons relating to the basic molecular structure of water, the majority of ice crystals – and almost all atmospheric ice crystals above temperatures of minus 80°c – are hexagonal in shape. The two basic shapes of hexagonal ice crystals are known as columns and plates; both these types of hexagons function as prisms, and the relative proportion and position of the two types explain a wide range of

A solar glory, Canada, 2005. Glories are only visible at a lower altitude than the observer, and their exact cause is not yet known, though an explanation based on the phenomenon of quantum tunnelling has been proposed.

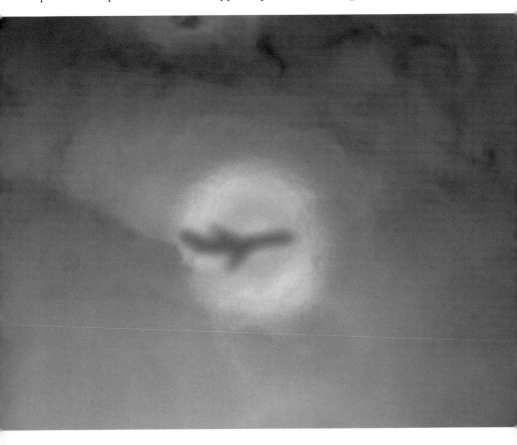

A 22-degree halo (top) and circumhorizontal arc (bottom).

atmospheric effects. One of these is the halo or nimbus, a circle with a radius of 22 degrees – red on the inside and blue on the outside – that can form around either the Sun or Moon. Halo production is driven by plate-shaped hexagonal ice crystals oriented with their flat faces pointing upwards and downwards, ideally within 1 degree of perfectly horizontal, into which sunlight passes through one vertical face and out through another.[16] (Halos should not be confused with coronas, which are smaller in diameter, brighter in colour and caused by water droplets.) Parhelia, also known as 'sun dogs' and 'mock suns', are bright, round or oblong rainbow-coloured patches that also appear 22 degrees on one or both sides of the Sun, but always at the same elevation above the horizon as the Sun is at any given moment and with their red sides nearest it. Here, the plate-shaped hexagonal crystals are again flat-side downwards, and the height of the parhelion is governed by their stability, with more stable crystals resulting in parhelia that are rounder. Halos and parhelia are seen together fairly often, as in the photo overleaf taken at the South Pole in January 2009.

The common term for circumhorizontal arcs, 'fire rainbows', is particularly unapt, in that they, too, are formed by sunlight passing into the side of a horizontally oriented field of hexagonal

Halo above Holy
Trinity Church,
Stratford-upon-Avon,
2014.

Full-circle solar halo with parhelia, South Pole, January 2009.

ice crystals – in this case, in through the sides and out through the bottoms. Circumhorizontal arcs are always roughly parallel to the horizon, and appear much more often in the mid-latitudes (for example, in the Mediterranean and in the American South) than in northern Europe or at the equator.[17] Even more confusingly, the term 'fire rainbows' is also sometimes applied to iridescent clouds, which are formed by wave-interference rather than refraction (through either water or ice), and have a wider array of different shapes. As well as the different processes and atmospheric media that produce them, both types of so-called fire rainbows appear to change colour as the observer moves, further setting them apart from true rainbows.

Column-shaped hexagonal ice crystals with a horizontal orientation can result in infralateral and supralateral arcs, if sunlight passes between one of their hexagonal ends and one of their rectangular sides. These rare arcs 'change their shapes dramatically with changes in solar altitude' and cannot be seen at all if the Sun

So-called fire rainbow.

is above 32 degrees.[18] Lastly, circumzenithal or Bravais arcs, which look like blurry, upside-down rainbows, are again caused by ice in cirrus clouds at altitudes of between 5,500 m and 12,000 m (18,000 and 40,000 ft) – in this case, by plate-shaped crystals.[19] They can generally only be seen in still air with the Sun at least 5 but less than 33 degrees above the horizon.[20]

Far from being simple or complete, the story of how we arrived at our current understandings of why rainbows occur when, where and how they do has been a long and twisting one: full of false starts, dead ends and occasional flashes of madness, alongside the brilliant insights and dogged experimentation that in the end yielded the answers that are – for the time being – considered correct. Our next chapter tells that story.

2 Rainbows in the History of Scientific Enquiry

We use the classical theory on Mondays, Wednesdays and Fridays and the quantum theory on Tuesdays, Thursdays and Saturdays.
Sir William Bragg[1]

The exact origin of what might be termed a scientific approach to the natural world remains a matter of some debate, amid competing claims that the scientific method originated in ancient Egypt or in seventeenth-century Europe or at various points in between. In any case, the history of the study of rainbows as scientific phenomena closely parallels the history of science itself. Dating from the fourth century BCE, Aristotle's rainbow theory owed something to earlier authors, especially Anaxagoras (d. 428 BCE), Anaximenes of Miletus (*fl.* 550 BCE) and Hesiod (*fl.* 700 BCE).[2] But in its level of detail and its power over subsequent generations, it was quite unparalleled, despite fatal flaws stemming from the fact that 'avoidance of experiment' lay 'at the heart of Aristotle's methodology on the rainbow'.[3] The history of the bow as an object of scientific enquiry is therefore largely the story of the acceptance, loss, rediscovery, re-acceptance and gradual abandonment of Aristotelian ideas over a period of two millennia. In comparison to this slow yet fascinating trudge from deference to self-confidence, the arrival of the 'modern' rainbow in the seventeenth century can seem like a sudden explosion, since which improvements to our understanding have again been relatively incremental and slow. While it is not feasible in the space of this chapter to provide a complete narrative of all the pre-modern rainbow theories and experiments about which anything is known, it is certainly possible to trace some of the most important turning points, and dead ends, in the study of this extraordinary phenomenon.

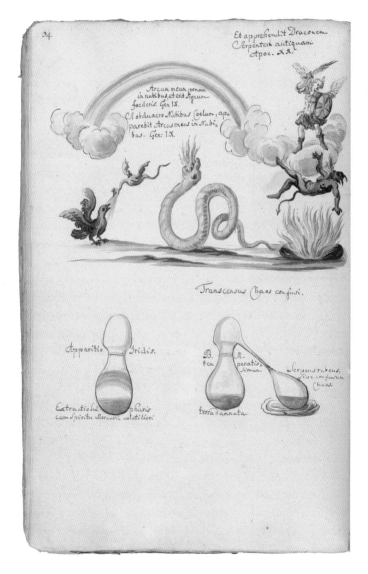

Rainbow, bottled rainbow, dragons, demon and warrior from the *Thesaurus thesaurorum et secretum secretissimum in quo omnia Mundi arcana latent*, a European alchemical text of *c.* 1725.

Aristotle: a half-baked solution becomes a new problem

Aristotle, who lived in the fourth century BCE, correctly understood several aspects of the rainbow: that it could not appear without the presence of water droplets in the atmosphere; that there were 'geometrical uniformities' in the relative positions of the bow, the observer and the Sun; and that the bow's shape might also be accounted for geometrically.[4] His belief that the bow consisted of only three colours – red, green and violet – was not necessarily wrong in itself, as such judgements are to some extent culturally based. However, other aspects of his explanation had manifest problems – for instance, his conviction that each of these three colours was a mixture (in different proportions) of white and black, with red containing the most white of the three, and violet the least. At any rate, these were the real colours; Aristotle saw yellow in the rainbow, too, but considered this to

Woodcut of Aristotle pointing to the heavens, in a 1496 French edition of his *Physics*.

be a mere trick of perception.[5] That there *should* be exactly three colours was given a variety of non-scientific or pseudoscientific justifications, including the division of events into three parts (beginnings, middles and ends), the division of space into three dimensions and even the use of the number three 'in the worship of the gods'.[6] Aristotle further held that the red band of the rainbow was the fattest in primary rainbows and thinnest in secondary ones. While aware of the existence of refraction – and even that this process might be 'effective' in the production of colour – he never suggested that rainbows were caused by anything other than reflection.[7] Violating his own dictum that the Sun was larger than the Earth, Aristotle also used reflection as a broad explanation for why daylight can be seen everywhere and not merely in 'that spot on which the sun's rays directly fall'.[8] He recognized that reflection operated according to natural laws of some sort, but expressed little or no interest in what these laws might be; and if Euclid's law of reflection was known during Aristotle's lifetime, Aristotle either rejected it out of hand, or was wholly unaware of it.[9]

Because he also supposed that every reflection represented a drastic weakening of the original image – as in the primitive metal mirrors with which he was familiar – Aristotle held that no tertiary rainbow could exist; following their exertions in creating the secondary bow, the reflections required to create a third would be too weak.[10] (It did not bother most of Aristotle's followers that if the

An imagined conversation between Aristotle, Ptolemy and Copernicus, engraved in 1632 by Stefano della Bella (1610–1664).

secondary bow were in fact merely a reflection of the primary, it would be U-shaped and not a second, larger arch.)

The apparent strength of the red band in the secondary bow, Aristotle said, was because this band, being lowest relative to the ground, is nearest to the observer.[11] However, this explanation was subverted by the fact that the red band in the primary bow was, by Aristotle's reckoning, the farthest from the observer yet also the brightest. So in the case of the primary bow, he explained this away using a separate argument that the redness/brightness of the outermost/highest band was due to it having the greatest overall area, much as the outside lane on a race course is the longest, and that it therefore had the most opportunity for reflection.[12] But obviously, these two explanations cannot be reconciled. Nor can Aristotle have spent much time observing rainbows in the mountains, since he maintained that 'the rainbow never forms a full circle, nor any segment greater than a semicircle'.[13]

Most importantly, in terms of his blockage of correct understanding of the rainbow by generations of later observers, Aristotle maintained that it was always reflected from clouds. This was an echo of the work of Anaximenes, who, two centuries earlier, had suggested that rainbows consisted of 'sunlight . . . returned to an observer from an impenetrable cloud'.[14] For Aristotle, however, the water droplets that gave rise to the rainbow did so because, though flat like household mirrors rather than spherical or spheroid, they were far too small to reflect the Sun as it actually appears, and in place of the Sun's image could reflect colours only. And crucially, Aristotle's clouds (and Sun, and everything else located in what we moderns would call both the sky and outer space) are attached to the inside of an imaginary sphere with the observer at its centre, and are thus necessarily concave.

The biggest of many problems with a rainbow based entirely on reflection is that it ought to grow taller as the Sun rises and shrink as it sets, rather than rise as the Sun sets and vice versa, which rainbows always do in reality. Another key implication of Aristotle's ideas was that the rainbow was exactly the same distance from the observer as the observer was from the Sun, regardless of whether the observer moved; and one wonders if this troubled even his ancient contemporaries, despite them being accustomed to thinking of the sky as the flat Earth's domed roof.[15]

The ancients after Aristotle

Though Posidonius took issue with minor aspects of Aristotle's rainbow in the second century BCE, the first major challenge came from Seneca the Younger (d. 65 CE), who hacked at the roots of the Aristotelian shrub by attacking its source material: the rainbow theory of Anaximenes.[16] If rainbows were indeed caused by some water particles in a cloud transmitting sunlight and others not transmitting it, Seneca asked, why was the bow not merely of two colours, light and dark? While at first tentatively accepting the Aristotelian view that the shape of the rainbow was produced 'by a cloud formed like a concave, round mirror', Seneca expressed nagging doubts about the fact that rainbows resembled the Sun in neither colour nor size.[17] He also moved 'acutely to deny from the outset that a rainbow can be formed by the reflection of the sun *in cloud*' on the grounds that clouds do not contain any raindrops, only matter from which raindrops are later produced.[18] Moreover,

> it is obvious that in the case of a rainbow no actual color is formed, but only the appearance of illusory color, the sort which . . . the neck of a pigeon takes on or gives up whenever it changes position. This is also the case in a mirror, which assumes no actual color, but only a kind of copy of the color of something else.[19]

Alexander of Aphrodisias (*fl.* 200 CE) was the pre-eminent ancient Greek commentator on Aristotle's works. He raised a question that cut right through the master's problematic dual explanation for the primary bow being brightest in the top band and the secondary bow brightest at the bottom: why is the

Frontispiece to the 1549 Venetian edition of the *Commentaria* of Alexander of Aphrodisias.

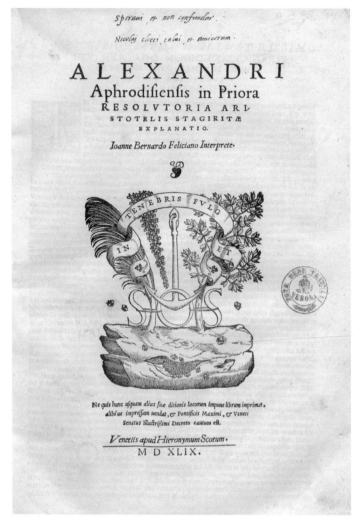

Speram, et non confundar.

Nicolai elicei enini, et amicorum.

ALEXANDRI

Aphrodifienfis in Priora

RESOLVTORIA ARI-
STOTELIS STAGIRITÆ
EXPLANATIO.

Ioanne Bernardo Feliciano Interprete.

Ne quis hunc ufquam alius fuæ ditionis locorum impune librum imprimat, alibi'ue impreffum uendat, et Pontificis Maximi, et Veneti Senatus Illuftriffimi Decreto cautum eft.

Venetiis apud Hieronymum Scotum.

M D XLIX.

whole area between the primary and secondary bows not also red? Indeed, why does it appear darker than either the area inside the primary bow, or the area outside the secondary bow? This darker inter-bow area is still known today as Alexander's band. But such a challenge was exceptional. Olympiodorus the Younger provided a staunch, programmatic defence of the Aristotelian theory, and authors in the West mostly followed his example until after the fall of Rome, when knowledge of the

Greek language (and with it, direct knowledge of what Aristotle had said) rapidly faded away.[20]

By the seventh century, despite the appeal of arguments that a three-colour rainbow accorded well with the Holy Trinity, a number of Christian thinkers had decided that the rainbow contained four colours corresponding to the four elements. However, they disagreed over which colours represented each of them. Saint Isidore (d. 636), for instance, saw a red, purple, black and white bow representing fire, water, earth and air, whereas the Venerable Bede (d. 735) saw the same elements as producing the rainbow colours red, blue, green and brown. But for the time being, European investigation of the bow's other aspects had effectively ceased.[21]

Clear image of Alexander's Band, the dark inter-bow area named for Alexander of Aphrodisias, who deployed it as part of an important early challenge to Aristotle's deeply flawed rainbow theory.

Rainbow scholarship in the Islamic world

King Ptolemy II of Egypt (r. 283–46 BCE) had purchased a great quantity of Aristotle's books, and this – along with the ongoing use of Greek in the region – may have helped prop up the Aristotelian rainbow theory in the Eastern Mediterranean.[22] But it was the emergence of Islam in the seventh century CE that gave new impetus to the study of Aristotle's rainbow writings, among many other ancient Greek scientific texts. In the 820s, the Syrian scholar Ayyub al-Ruhawi – better known as Job of Edessa – provided the first serious departure from Aristotle since Alexander of Aphrodisias. Al-Ruhawi did not dispute that the sky was domed, or specifically that it 'appears to us like a vault', but he jettisoned Aristotle's demonstrably false assumption that rainbows are only ever seen against a background of clouds.[23] He still maintained that they were caused by reflection only, but reflection from 'thin' and 'thick' forms of humidity in the air – even air that appears 'clear' to the naked eye. Like Aristotle, Al-Ruhawi saw the rainbow as being of three colours, but in his case, yellow replaced violet; and to explain why three different colours were produced, he again relied on fairly vague references to humidity.[24]

This was merely a small down payment, however, on the advancements provided by later scholars in the Islamic world. The Persian Ibn-Sīnā (d. 1037), known in the West as Avicenna, rejected the cloud-mirror concept on the grounds of his own observations, and further argued that water drops in the atmosphere were 'a *necessary*, rather than merely a *sufficient*, condition for the rainbow'.[25] And while he accepted that the bow was of the same three colours assigned to it by Aristotle, Ibn-Sīnā was deeply troubled by the reversal of colours in the primary and secondary bows, which he felt that neither Aristotle's theories nor his own could do anything to explain.

It fell to Ibn-Sīnā's Basra-born contemporary Ibn al-Haytham ('Alhazen', d. 1039) to bring the notion of refraction to the problem. Through carefully designed experiments involving water-filled glass globes, al-Haytham established the first mathematical rules for the refractive bending of light. Though these laws were not perfectly accurate, and he did not apply them to his fundamentally Aristotelian explanation of the rainbow, they laid the groundwork for further experimentation by similar methods. In particular, Qutb al-Din al-Shirazi (1236–1311) 'at a stroke laid bare the essence of the rainbow's optics' – namely, that it is caused by two refractions and one reflection of light within a raindrop.[26] In Europe, unfortunately, Qutb al-Din's unprecedented insight barely became known; and even if it had been, it would have had to compete head-to-head against Aristotle's original theory, newly rediscovered.

Conventional portrait of Ibn-Sīnā (Avicenna) on a silver vase in Iran.

The Western rediscovery of Aristotle

With the Spanish conquest of the city of Toledo in the late eleventh century, a great library containing the works of Aristotle, Euclid and Alexander of Aphrodisias, as well as Islamic authors including Ibn-Sīnā, suddenly became available to the Christian West. The supply of newly translated scientific texts was matched by demand for curricula from the wave of European universities that were founded from the mid-twelfth century onwards. As in so many other areas of science, the medieval universities' nearly universal support for Aristotle would have a great deal to answer for in the sphere of rainbow theory. However, a University of Paris graduate working in Oxford, Robert Grosseteste (Bishop of Lincoln from 1235; d. 1253), was the first Westerner to work out a rainbow theory involving refraction. Proceeding from the problem that a rainbow produced by reflection would rise and fall with the Sun, rather than inversely to it, he proposed that refraction must instead be the key to the process of rainbow production. And, in perhaps the strongest departure from Aristotle up to that time, he assigned the inception of his series of three refractions to a mostly round cloud *behind* the observer. However, in a convoluted description that would have troubled even the writers of *Doctor Who*, Grosseteste's refracting cloud had to be concave in some places and convex in others, and also able to emit an invisible 'pyramid or

cone' of dense moisture.[27] Modern experts disagree vigorously as to whether this cone was meant to point steeply upwards from the ground into a round cloud, or downwards to the centre of the Earth from a cloud shaped like an inverted bowl.[28] In any case, somewhat like Aristotle's rainbow, Grosseteste's was 'projected . . . as though on a screen' onto a (second) cloud.[29] But in contrast to Aristotle's, this receiving cloud was also externally convex, not concave: in other words, the ancients' dome-of-heaven concept did not enter into the bishop's explanation at all. It is frankly astonishing that people had believed for so long that all rainbows were projected onto (or perhaps from) the literal edge of the world. For, as Philip Fisher rather poetically puts it,

> The rainbow was always . . . the one part of heaven that
> occurred here on earth, just over there, touching a neighbor's

Woodcut scene illustrating various properties or 'feats' of light, from a 1572 edition of the *Opticae thesaurus* of Ibn al-Haytham (Alhazen).

Thirteenth-century optical experiment utilizing a water-filled glass sphere, variously attributed to Roger Bacon and to Bacon's teacher, Robert Grosseteste.

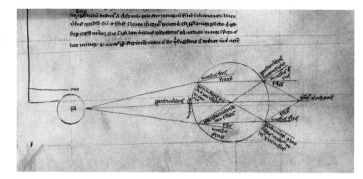

barn and that familiar tree, in front of the hill, closer than the nearest town.[30]

And yet, as we shall see, this idea did persist – in some circles, for centuries after Grosseteste's death.

Despite his precocious appreciation that refraction had a role in rainbow production, Grosseteste was 'concerned primarily with the shape of the bow', and left only a 'hint' that he might have associated refraction with the rainbow's colours.[31] Moreover, in abandoning the Aristotelian dome-of-heaven idea, or at any rate the notion that rainbows appear on the interior surface of that dome, Grosseteste also jettisoned the only extant explanation for the fact that rainbows move as the observer moves – namely, that they are exactly as far away as the Sun. Though tortuous, 'crude' and ultimately quite wrong, Grosseteste's rainbow theory must nevertheless be counted as an important step in the right direction, for it was arguably the first to challenge Aristotle fundamentally, and certainly the first refraction-based explanation of the bow to be presented to a wide European audience.[32]

Grosseteste's pyramid-of-moisture idea was adopted – some would say stolen – as part of the 'correct' explanation of the rainbow in a voluminous history of rainbow theories compiled by Albertus Magnus (d. 1280). It was also accepted by John Peckham, Archbishop of Canterbury (d. 1292). But the credit for Grosseteste's real insight, that refraction played a major role, was, until fairly recently, assigned to the Polish-German scholar Witelo of

Viterbo (*fl.* 1270), an expert on the works of al-Haytham.[33] Witelo's genuinely original contributions were twofold. First, he suggested – much like his exact contemporary Qutb al-Din, but to a European audience – that refraction and reflection were both involved in the rainbow's production. This particular part of Witelo's work was far less satisfactory than Qutb's, however, in that Witelo postulated an interplay between multiple raindrops, in which the refractions were internal to the drops, but the reflections external. Second, Witelo insisted that refraction be studied via 'experiments with instruments' rather than the suppositions and analogies that previous scholars had tended to rely upon.[34]

Witelo's main rival and nearest European contemporary, Roger Bacon (d. 1292), had been a student of Grosseteste's at Oxford and has sometimes been erroneously given credit for the authorship of Grosseteste's *De iride*.[35] Bacon is often derided for his rejection of the role of refraction and defence of a reflection-only rainbow. Other aspects of his work on the subject have also been critiqued as 'thoroughly "medieval" in the popular sense of the word': for instance, he expanded the number of rainbow colours to five – red, blue, green, black and white – but on the grounds that the human eye contained 'five bodies . . . three humors and two coatings'.[36] Nevertheless, Bacon made four significant contributions to the growing debate. First, 'he rejected refraction for good reasons, by advancing sound criticisms against some of the more absurd aspects of Grosseteste's refraction theory'.[37] For instance, the appearance of rainbows in small, localized sprays of water ruled out Grosseteste's triple refraction starting from an extraordinary cloud-cone, which could hardly exist indoors – if, indeed, it could exist at all. Second, Bacon used direct observation to establish that the maximum elevation of the bow was 42 degrees: a figure that was very close to correct and would not be bettered for nearly four centuries.[38] Third, he pointed out that

if two people stand observing the rainbow in the north and one moves westward, the rainbow will move parallel

VITELLIONIS MA‑
THEMATICI DOCTISSIMI περὶ ὀπτικῆς,
id est de natura, ratione, & proiectione radiorum uisus, lu‑
minum, colorum atq; formarum, quam uul‑
go Perspectiuam uocant,
LIBRI X.

Habes in hoc opere, Candide Lector, quum magnum numerum Geometricorum elementorum, quæ in Euclide nusquã extant, tum uero de proiectione, infractione, & refractione radior; uisus, luminum, colorum, & formarum, in corporibus transparenti‑ bus atq; speculis, planis, sphæricis, columnaribus, pyramidalibus, côcauis & conuexis, scilicet cur quædam imagines rerum uisarû æquales, quædã maiores, quædam minores, quædam rectas, quædã inuersas, quædam intra, quædã uero extra se in âere magno mi‑ raculo pendentes: quædam motum rei uerum, quædã eundem in contrariû offendant: quædã Soli opposita, uehementissime adurant, ignemcq; admota materia excitent: deq; umbris, ac uarijs circa uisum deceptionibus, l quibus magna pars Magiæ naturalis de‑ pendet, Omnia ab hoc Autore (qui eruditorum omniû consensu, primas in hoc scripti genere tenet) diligentissime tradita, ad solidam abstrusarum rerum cognitionem, non minus utilia & iucunda.) Nunc primum opera Mathematicoq; præstantiss. dd. Ge‑ orgij Tanstetter & Petri Apiani in lucem ædita.

Norimbergæ apud Io. Petreium, Anno MDXXXV.

The 1535 engraved title page of *De natura* by Witelo of Viterbo, one of the first scholars to suggest that refraction and reflection were both involved in the rainbow's production, and that the former should be studied via formal experiments.

to him; if the other observer moves eastward, the rainbow will move parallel to him; and if he stands still, the rainbow will remain stationary. It is evident, therefore, that there are as many rainbows as observers ... and each observer must see his own rainbow.[39]

As well as being quite true, this is of course fundamentally at odds with the idea that the bow is projected as if on a screen, which had been promoted by Grosseteste and a high proportion of his predecessors.[40] Finally, Bacon 'called attention to the function fulfil‑ led by individual drops, arguing that the rainbow appears in a different set of drops for each observer'.[41] But a key question, as D. C. Lindberg reminds us, remained unanswered:

Why does the rainbow appear as an arc and not a full circle (or semicircle) of color? ... This is one phenomenon that neither a simple reflection theory nor a simple refraction theory can satisfactorily explain.[42]

The 'modern' theory from Theodoric to Descartes

In or before 1304, inspired chiefly by the works of al-Haytham, a German Dominican cleric and former University of Paris student named Theodoric of Freiberg (d. *c.* 1310) conducted a series of experiments involving the use of water-filled spheres of glass to simulate raindrops. While his methods in and of them‑ selves were 'particularly pregnant for the future', his results

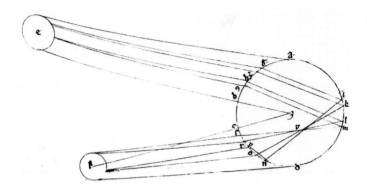

Ray-tracing from
Theodoric of
Freiberg's *De iride*
(*c.* 1305).

allowed him to argue correctly that the rainbow colours red, yellow, green and blue always occur in this 'inviolable order'.[43] Moreover, he established valid reasons for the relative positions of the primary and secondary rainbows; for why Alexander's band was darker than the areas outside the two bows; for why the colours of the secondary bow occur in reverse order; and most significantly, for how the two bows were produced via a particular number and sequence of refractions and reflections. 'Gone is the need for a problematic reflecting (or refracting) cloud – each raindrop is a self-contained prism *and* mirror.'[44] His initial decision to focus his enquiry on the raindrop may have been partly inspired by the work of Roger Bacon, but in any case, Theodoric was a prolific scholar who 'probably was familiar with all major thirteenth-century writers on the rainbow', with the exception of Qutb al-Din.[45]

As correct as he was about the rainbow in many respects, however, Theodoric was wildly wrong in others. In contrast to the nearly correct measurements provided by Bacon, Theodoric measured the primary bow's height as just 22 degrees, and that of the secondary bow as 33 degrees. Unlike Bishop Grosseteste, Theodoric implicitly accepted the Aristotelian dome-of-heaven idea – and even went beyond it, in the sense that his own model of rainfall implied the drops falling not straight to the ground, but along the same curvature as the world's 'roof', or what he referred to as the 'altitude circle'. He held that the rainbow contained only the four colours red, yellow, green and blue, and also seemed to think that rays of sunlight are not parallel to each other, but spread out from the Sun in an array, perhaps implying that the Earth is larger than the Sun. None of this, however, should detract from the fact that Theodoric's work on the production of rainbows was substantially correct in its 'qualitative insights' and faulty only in its 'quantitative details'.[46]

It is widely reported that Theodoric's breakthroughs were swiftly and tragically forgotten for five centuries, until their discovery and publication by Giambatista Venturi shocked the world in 1814.[47] This particular myth accords nicely with the 'naïve' but 'widely held' idea that 'the first theory of the rainbow was given by Descartes in 1637'.[48] In fact, the years between 1500 and 1700 saw more books on the rainbow published than in all time before or since, though in fairness, many of these were fundamentally non-scientific works devoted to prognostication, alchemy and the like, even where men of science were credited with their authorship.[49] Certainly, the sixteenth and seventeenth centuries saw determined elaborations of the Aristotelian reflection-only rainbow by, among others, Francesco Maurolico (1494–1575). Attacking Peckham and Witelo for their 'obscurity', Maurolico proposed a theory which, though right in

This diagram from *De iride* clearly shows how the traditional 'dome of heaven' concept added a needless layer of complexity to Theodoric's rainbow theory.

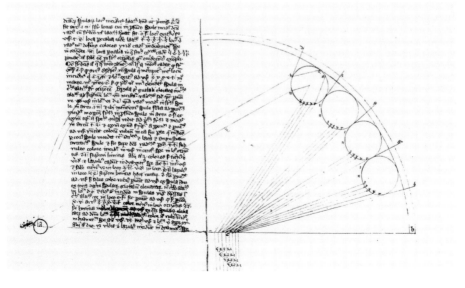

a few respects, fundamentally required sunlight to have 'terminal points' on an imaginary plane standing perpendicular to the surface of the Earth.[50] In effect, he had replaced the ancients' inverted bowl of heaven with an inverted box. Yet the idea that Theodoric's rainbow theory had been lost is not quite correct, insofar as it was evidently known to Themo Judaei (*fl.* 1360), Regiomontanus (1436–1476) and Jodocus Trutfetter (*fl.* 1514), and it may even have been taught at the University of Erfurt in Trutfetter's time and beyond.[51] Marco Antonio de Dominis (1560–1624), renegade Catholic Archbishop of Spalatro – and later, Protestant Dean of Windsor – mysteriously 'gave a rather better explanation of the primary rainbow than any published before 1637, when Descartes gave the correct elementary theory of both bows'.[52] But de Dominis' contribution of 1611 was mysterious only to the extent that we believe Theodoric's rainbow theory had been definitively lost by that date, for we now recognize de Dominis' work as being an only 'somewhat distorted' version of Theodoric's.[53] Its greatest flaw was that it 'overlooked the fact that the rays must be refracted on emerging from the raindrops, as well as upon entering'.[54]

Johannes Kepler, who associated each of the six known planets with a particular musical scale (as shown), also compared the colours of the rainbow to 'the infinity of tones in the musical octave'.

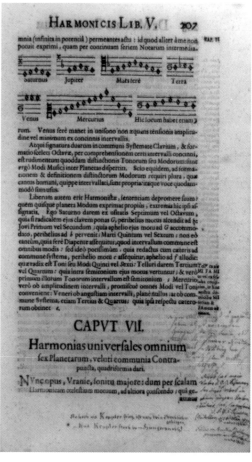

Johannes Kepler (1571–1630) came at the rainbow from a decidedly oblique direction: a quest to apply the harmonies of music and mathematics to the mystery of colours. Perhaps the most remarkable aspect of his views on the rainbow, however, was how much they changed over time. Writing around 1599, Kepler compared the colours of the rainbow to 'the infinity of tones in the musical octave', with yellow operating as 'a sort of mean'. The darkening of the rainbow from yellow via red into black was caused by 'crass material in the cloud', while refraction was responsible for the bow's brightening from yellow via green, blue, purple and violet.[55] It is clear that Kepler initially followed pre-Aristotelian views that a cloud located in front of the observer – but acting as a whole, and not via its individual particles of moisture – was responsible for the phenomenon. Also, Kepler's 'refraction' at this early period may have been merely a reproduction of the ancients' confusing but commonplace use of 'reflection' and 'refraction' as synonyms. By 1604, however, Kepler had changed his tune completely, and was supporting an extreme version of Grosseteste's view that the rainbow was caused by light refracted through a cloud placed between the Sun and the observer's back.[56] It was only after this that Kepler first resorted to model-raindrop experiments, allowing him to conclude that colour 'seems to depend upon the magnitude of the angle of incidence', and struck up a correspondence with the Oxford mathematician Thomas Harriot (1560–1621), who had established – but not publicized – the law of refraction between 1597 and 1606.[57] Even following this interchange of ideas with Harriot, however, Kepler's work on refraction and reflection within the raindrop was hampered by an assumption – shared with many of his predecessors, ancient and modern – that the primary rainbow's radius was 45 degrees, not 42.[58] Never realizing that this was the problem, he ascribed the errors in his calculations of refracted rays to variations in water temperature. Though Kepler was right about refraction decreasing as water temperature increases, this effect is far too small to account for the discrepancies in his research.

As previously mentioned, René Descartes (1596–1650) is perhaps the person most widely credited with cracking the secrets of the rainbow, even if no one would nowadays dare to suggest he was the first person to try. In spite of or because of this primacy, however, 'few scientists have been subjected to charges of plagiarism more frequently'.[59] Though 'it is quite probable that he had never even seen' the work of de Dominis, Descartes was accused of plagiarizing from him by Gottfried Leibniz (1646–1716) and Isaac Newton (1642–1727), among many others; and the opinion that Descartes 'was unjust in not acknowledging' his great intellectual debts to de Dominis was 'widely held' as recently as the mid-twentieth century.[60] But Carl Boyer, the pre-eminent historian of rainbow theory and experimentation, concluded in 1952 that

> Descartes believed, mistakenly, that he was the first one to study the rainbow through experiments with a large spherical globe of water which served as a magnified raindrop.[61]

Descartes also may have independently arrived at the correct law of refraction, though this had been taught at the University of Leiden by Willebrord Snellius (1580–1626), who died about two years before Descartes' rainbow work commenced. The key difference here between Snellius and Descartes was that the latter *knew* 'that he had discovered a completely general and powerful description of the path of light rays within transparent media'. Neither man, however, thought to link the precise angle of a light ray's deviation to its colour.[62]

Though largely 'indifferent' to hands-on experimentation, Descartes did not 'entirely reject' it either, and in the case of the rainbow he performed 'endless' observations.[63] Following the voyages of Columbus and Magellan, the dome of heaven was no longer seriously believed in, except by 'illiterate persons'.[64] But within the field of rainbow research, Descartes' experiments finally put paid to it once and for all: by showing that only the angle of observation, and not the linear distance between the raindrops and the observer, determines what rainbow colours he can see.[65] To all of the correct aspects of Theodoric's theory, Descartes added the crucial insight that there is a particular point near the top of a spherical water drop, at and above which all rays entering the drop (via the correct sequence of refraction, reflection and refraction) become concentrated at roughly the same angle as one another when they exit the drop's lower hemisphere – that is, head back in the general direction of the Sun and the observer, but now *nearly* rather than *exactly* parallel, in a tight bundle 40 to 42 degrees below the level at which the sunlight first entered the drop. Bluntly put, 'this concentration of sunlight at exit angles near 41 degrees produces rainbows.'[66] Descartes also came to realize that the other side of the coin of this concentration process is that the sky is starved of light within the 9-degree inter-rainbow area known as Alexander's band.

Ancient and medieval authors had been aware of prisms, and even conducted experiments with them, but tended to reject the relationship of the spectrum to the rainbow as apparent rather than actual. This was due to a widespread belief that rainbows were mere optical illusions, whereas prismatic spectra could be projected onto walls, approached and touched, and were therefore real – though perhaps not *as* real as the colours of dyed cloth. Descartes rejected such distinctions. Because prisms do not operate either via curved surfaces or reflection, he was also able correctly to surmise that the rainbow's colours were produced by some other aspect of the rainbow-formation process. From this point, however, his colour theory began to fall apart quite rapidly. Briefly put, Descartes came to believe that colours were caused by microscopic globes of air, struck initially by the force emanating from a light source and then rolling onwards in a great mass and striking each other like billiard balls; the forward travel of the balls transmitted light *per se*, while their differing rates of spin produced colours. Moreover, in seeming contradiction to this (itself quite wrong and strange) theory, Descartes' work also wrongly implied that light travels faster in water or glass than it does in air – in reality it slows down by nearly 25 per cent in water and around 33 per cent in glass – and this notion 'hung on stubbornly for more than two centuries'.[67] Unsurprisingly, his work on the rainbow could not be used to explain its colour-order.

O

P

O

Fig I. B A C

114 PRINCIPIORUM PHILOSOPHIA

No-one abhorred a vacuum quite as much as René Descartes, for whom even outer space was made up of solids in collision. From the 1644 Amsterdam edition of *Principia philosophiae*.

'Formation of the Rainbow', illustration from the *Physica sacra*, by Johann Jakob Scheuchzer (1672–1733).

Newton and beyond

Right though much of it was, Descartes' work on rainbows was not immediately accepted as such even by a large minority of his fellow European scientists, for whom 'the Aristotelian rainbow remained a fixture of the intellectual landscape.'[68] Both Pierre Gassendi (1592–1655) and Francesco Grimaldi (1618–1663) were rare exceptions proving the rule; but by and large, Descartes was not generally accepted as the principal authority on the subject until the 1680s, some thirty years after his death.[69] Descartes' colour theory, which was obviously unsatisfactory even when newly written, was quickly supplanted by the theory of Edme Mariotte (1620–1684), who held – also quite incorrectly – that white light operated under wholly different laws of refraction than those that governed coloured light.

The problem of colours was also being approached from a more practical perspective – namely, the desire to eliminate colour distortions from new devices including the first practical microscopes (invented *c.* 1590) and refracting telescopes (1608). It was from this direction that Newton first embarked on his 'singular mental voyage'.[70] Aware of Descartes' theory of the rainbow by 1666, when he was aged just 23, Newton had already had the insight that the process of refraction might be colour-dependent; but he later conducted a dramatic series of experiments that demonstrated this conclusively.[71] A prism was placed over a hole in the wall of a building, and natural sunlight shining through the hole was projected onto the interior wall opposite. Newton found that the image of the Sun thus projected was always an oblong, four or five times longer

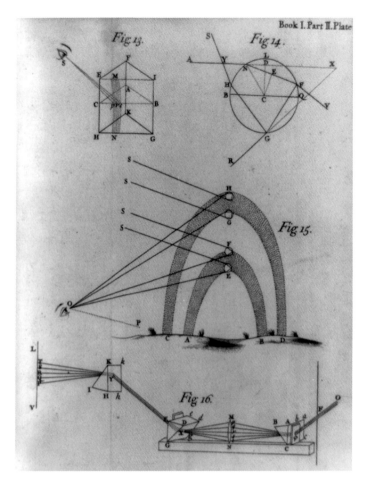

Rainbow and prism diagrams from Isaac Newton's *Opticks* (1704).

than it was wide – regardless of the position of the prism, the position of the Sun or the size of the hole. Because nothing he did in this scenario could make the Sun appear round, Newton realized that his initial insight had been correct: angles of refraction vary by colour. He later repeated the experiment with the planet Venus replacing the Sun, and achieved much the same results.[72]

Initially, Newton believed that there were five main rainbow colours – red, yellow, green, blue and purple – separated by four or more 'intermediate ones'.[73] Eventually, through painstaking experimentation, he would arrive at the seven 'canonical' colours still generally accepted today, though these were considered 'heretical' by many in his own time.[74] In part, this hostile response was because Newton had suggested – correctly – that all of the rainbow colours were present in white light, and not the result of mixing light with darkness in differing proportions, as had been generally believed from at least the time of Aristotle onwards.

To Roger Bacon's key realization that no two people see the same rainbow, Newton added the idea that it is impossible for 'any one perceiver [to] see the identical rainbow from moment to moment', further focusing our attention on 'the crucial completing role of the viewer in achieving the rainbow image'.[75] Newton also gave the most accurate measurements so far of both the primary and secondary bows – that is, that the former stretched from 40.283 to 42.033 degrees, and the latter from 50.950 to 54.117 degrees. Moreover, he recognized that it would be possible to arrive at a general law of the dispersion of light, which, if joined to the already discovered law of refraction, would allow one to 'completely specify refraction of colors between any two media'.[76] Unfortunately, this insight was not matched by any success (or indeed sustained interest) in actually establishing such a law. Yet, the main flaw in Newton's understanding of colours – which would nevertheless stand all but unchallenged through the eighteenth century – was that he conceived of light as a particle rather than a wave. And like Descartes, Newton assumed that light moved more quickly through denser substances.[77] It is hard to say if Newton's rejection of light as a 'pulse' and his insistence that it was 'corpuscular' was more an effect or a cause of his deep personal loathing for Robert Hooke (1635–1703), who had proposed a rudimentary wave theory of light in 1665.[78] But in a curious parallel to the thirteenth-century debate on the all-reflection versus all-refraction rainbow, neither an all-particle nor an all-wave theory of light would prove entirely satisfactory. By the twentieth century, both would give way to 'a new ontology' in which, as A. Lande explains,

> objective physical reality in the traditional sense is discounted
> as old-fashioned, at least in the microphysical domain where

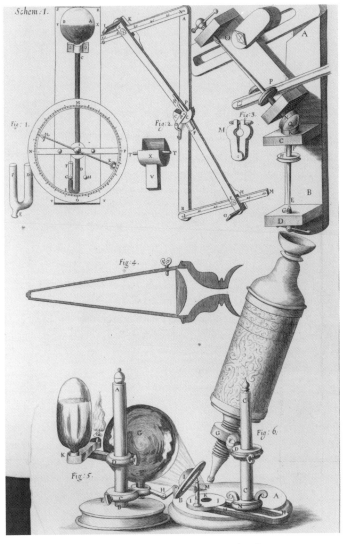

Though much better known for his contributions to microscopy, seen here in this page from his *Micrographia*, Robert Hooke proposed a rudimentary wave theory of light in 1665.

we can only form pictures of a subjective kind, even opposite ones as those of discrete particles and of continuous waves.[79]

With regard to the rainbow in particular, the modification of Newton's ideas was slow indeed, amid what can fairly be described as his eighteenth-century deification. It would fall to Thomas Young to begin the reconciliation of 'everyday perceptual

The rainbow surmounting a vista of mankind's actual and expected scientific and technical achievements. By Etienne Voysard (1746–1812) after Claude-Louis Desrais (1746–1816).

experience' to 'Newton's essentially correct physical model of light', a process that would be further refined by James Clerk Maxwell's (1831–1879) identification of light as waves of electro-magnetic radiation in 1865.[80] Because it was largely rejected by scientists, though accepted by artists, the colour theory published by J. W. von Goethe in 1810 will be addressed in Chapter Five, below.

Of course, nothing in the present chapter is intended to suggest that myriad other understandings of what rainbows were, how and why they formed, and what they meant had not been arrived at in other historical and cultural contexts. The next chapter will survey these alternatives.

3 Rainbows and Myth

Not unexpectedly, given its apparent immensity, brilliant colours and sudden appearances and disappearances, the rainbow has been assigned mythic or spiritual significance by a bewildering array of national, confessional and cultural groupings worldwide. There is no general agreement about whether certain common rainbow beliefs spread between, or were independently invented by, cultures separated by thousands of miles of desert, jungle or ocean. Rainbow-citing end-of-the-world myths, for example, occur in apparently disconnected traditional cultures in eastern and western Africa; the Baltic states; northern India; southern China; several Pacific islands; Bolivia; Peru; Argentina; Paraguay; the southeastern u.s.; and eastern, western and northern Canada. Many of these can be dismissed as the warped residue of long-forgotten Christian missionary activity that spread the story of Noah and the flood; others cannot. The end-of-the-world rainbow stories shared by the Toba, Mataco and Lengua peoples of the Gran Chaco region of South America, for instance, bear no resemblance to Judaeo-Christian ones, but a very close similarity to some New Guinean and other legends of the southwestern Pacific region.[1] While bearing firmly in mind that cross-cultural comparisons of myths 'can only lead to valid conclusions after intensive study of the belief or custom in each culture-area separately', the partial evidence assembled by anthropologists over the past century and a half is highly intriguing, as much for the similarities among global rainbow myths as for their many differences.[2] Certainly, the rainbow has until relatively recently been an almost universal object of veneration, awe and fear.

The rainbow as bow

The Judaeo-Christian significance of the rainbow is derived from the story of the flood in the Old Testament's Book of Genesis, after which God sets his bow in the clouds, 'for a token of a covenant, betweene me and the earth ... [that] the waters shall no more become a flood to destroy all flesh'.³ Many scholars identify this biblical event with a particular real flood that occurred around 2900 BCE in southern Iraq, the story of which reached the compilers of the Old Testament, working in the sixth to fourth centuries BCE, via earlier Babylonian and Assyrian traditions.⁴ In medieval Christianity, the rainbow became prevalent in artistic depictions of the Apocalypse, as well as in some of the most popular Corpus Christi mystery plays. With the advent of Protestantism in the early sixteenth century, God's bow almost immediately became associated with the radical Left – seemingly via a sense that the 'meek', broadly identified as the poor, would indeed inherit the Earth in the End Times, pursuant to the promise made by Christ in the Sermon on the Mount.⁵ However, amid increased scientific appreciation of the causes and forms of the rainbow, trailed only slightly by public acceptance of the latest scientific explanations, most Westerners soon lost their ability to relate to the rainbow as something other than an optical and meteorological phenomenon. One loss in

Rainbow of God's covenant with Noah in a J. & R. Lamb Studios stained-glass-window design from 1857.

Hans Memling, *St John Altarpiece*, *c.* 1479, oil on oak panel. Flames shoot from the rainbow surrounding the figure of God in the right-side panel.

particular was the sense that the rainbow is, or is representative of, a bow: a weapon, used for war or hunting, which had been invented in the upper Palaeolithic era, and arguably brought to a peak of perfection in the fifteenth century, on the eve of its supersession by handheld firearms.

Medieval Europe shared this perception of the rainbow-as-bow with much of the Near East and parts of India. The senior Hindu god Indra, also known as Meghavahana ('rider of the clouds'), is a weather-controller believed to push up the sky each day to release the personification of the dawn, known as Ushas, from a cave.[6] Indra's main weapons are a bow and *vajra*, the latter meaning both 'thunderbolt' and 'diamond' – and for some, 'the diamond shining like a rainbow', a fairly clear allusion to prismatic phenomena.[7] In the Sanskrit language, the rainbow is called *indradhanus*, 'the bow of Indra', from which also comes the Bashgali word for rainbow, *indron*.[8] Somewhat controversially, it has been asserted that Indra or a theoretical proto-Indra figure spread both eastwards and westwards: becoming Vajrapani Bodhisattva in the Mahayana tradition, the most prevalent form of Buddhism worldwide and the main form of that religion in China; and in the European context, the thunder gods Zeus (Greece), Jupiter (Rome), Thor (Scandinavia and Denmark), Perkunas or Perkuno (Latvia and Lithuania), Ukko or Uka (Finland and Estonia) and Perun (Russia).[9] In any case, by the

55

time he reached Japan under the name Agyo, this god was carrying not a bow but a club made of diamond; and his relationship to the rainbow, as distinct from the spectrum, had been broken.[10] Ukko, Perkunas and Perun, meanwhile, were all armed with bows, but again without their legends placing any special emphasis on rainbows among a host of weather phenomena that these deities were said to cause, both accidentally and on purpose.[11]

Another Hindu mythological figure associated with the rainbow-as-bow is the hero-king Rama, seventh of the ten avatars of the god Vishnu in the Vaisnava tradition. One Bengali word for rainbow, *rongdhonu*, literally means 'Rama's bow', though other words for rainbow in the same language include *dēbāyudha* ('a divine or celestial weapon') and *indrāyudha* ('the bow of Indra'), as well as simply the name *rama* by itself.[12] Lying geographically between the above-mentioned Indian and European traditions is the pre-Islamic weather-god Quzah, formerly worshipped in the Muzdalifah area of Arabia.[13] His bow (*qaws Quzah*) remains the nearly universal Arabic term for rainbow today, though some recent scholars have questioned this derivation.[14] It is also possible, though far from conclusively demonstrated, that the ancient Egyptians associated rainbows with war-bows, at least visually.[15] And medieval Englishmen were happy to note that the unvarying position of the rainbow – aimed away from Man, and at God – reflected the state of wary truce between the two that had prevailed since the time of Noah and the ark.[16] Though this particular mythic perspective has now vanished, the English language cannot describe a rainbow without reference to it: the 'bow-ness' of the bow intractably, implausibly, indestructibly remains.

Though the Judaeo-Christian rainbow was, or was representative of, a powerful weapon, it was a weapon that would never be used; and in the West, the residue of this peculiar and contradictory myth has been that rainbows are now almost inevitably regarded as benign. Likewise, in the minority of world cultures that have regarded the rainbow as a benign object rather than a malign being, it is very often seen as part of the personal kit of

Ukonsaari island in
Finland is sacred to
the thunder god Ukko.

an anthropomorphic deity. In this sense, the bow-rainbows of Jehovah, Indra, Ukko, Quzah and the rest can be related not merely to each other, but to the mythical belt-rainbows of Armenia, Albania and Livonia; the necklace-rainbow of Ishtar (the Assyrian, Babylonian and Akkadian goddess of war, sex and fertility); and even, perhaps, the severed penis-rainbow of the hermaphroditic Balinese goddess Uma.[17]

The rainbow as cosmic architecture

As will have been noted from the previous section, Indo-European belief in a bow-armed thunder-god appears to have stopped at the border between Finland and Sweden, to the west of which godly thunderbolts were believed to emanate from a war-hammer instead. The global fame of the pagan Scandinavian rainbow bridge known as Bifrost, Bilröst or Asbru has been rein-forced by generations of Marvel comic books and, in a twisted form, by the viral popularity of the anonymous poem 'The Rainbow Bridge' from before 1993, which promotes the belief that pets and their owners will be reunited in the afterlife. Such is the popularity of this pet-centred bridge that it is increasingly incorporated into mainstream religious practice, for instance, in Japanese Buddhism.[18] In actual Norse mythology, the bridge connects the mundane world (Midgard) to the godly realm (Asgard), which will be destroyed during Ragnarok (the 'twilight of the gods'), after which the world will be drowned in a great flood. Bifrost is explicitly described as a rainbow of three colours in the thirteenth-century *Prose Edda* by Snorri Sturluson, upon whose work much of our knowledge of earlier Scandinavian mythology crucially depends. Some modern scholars have speculated that the three colours in question may have been red, blue

Lord Rama with bow in a Dussehra Navratri festival poster from India.

Statue of the goddess
Uma, Cambodian,
8th century CE.

and green; but only the red is actually attested in the original tales.

Amid the perhaps undue prominence of this medieval Scandinavian tradition in modern culture, it is frequently forgotten that other myths of the rainbow as an inanimate architectural element of the universe are widely dispersed around the globe. Bifrost itself is conceivably a version of the rainbow pathway used by some versions of the Graeco-Roman messenger goddess Iris to move between godly and human realms. Iris, however, was for some ancient authors so closely identified with the rainbow that the 'bridge' concept disappeared; her name remains the common word for 'rainbow' in Latin, as well as the root of the Spanish and Portuguese words for the phenomenon, and of the English word 'iridescence'. Some European folk traditions also preserved into modern times an idea that the rainbow was not Iris' body but her scarf, an idea with which Isaac Newton was familiar, and which presumably inspired the animation of Iris' cloak in the 'Pastoral Symphony' segment of Walt Disney's *Fantasia* (1940).

Riddles involving the rainbow are extremely common in the folklore of Norway, Sweden, Denmark and northern Germany, though more of these liken the phenomenon to a piece of cloth than to a bridge – making a common origin in the Old Norse tales unlikely. This has led at least one scholar to argue that the rainbow-bridge Bifrost 'is not of Norwegian origin at all' but an interpretation of the Graeco-Roman goddess Iris that arrived in Scandinavia via Germany. In any case, riddles about rainbows are considerably rarer in European cultures than ones about 'other celestial phenomena, such as thunder, sun, moon, and stars'.[19]

Certain Indonesian, Japanese, Khmer, Siberian Buryat, Hawaiian, Tlingit and Hopi traditions also cite a rainbow bridge connecting the heavens with the Earth and/or the dead with the living.[20] The Chumash people of California, meanwhile, have a story that combines elements of rainbow-bridge mythology with the rainbow-snake mythology that is even more common around the Pacific Rim, as well as in Africa (see next section). A small number of other traditions around the world have perceived the rainbow as an architectural element other than a bridge: for example, a rafter (Navajo), or 'a kind of celestial girder' in the house of the Queen of Heaven (Zulu).[21] More prosaically, a certain type of real bridge in medieval China was called a 'rainbow', after its shape.

Friedrich Wilhelm Heine (1845–1921), *Battle of the Doomed Gods*, illustration from Wilhelm Wagner's *Nordisch-Germanische Götter und Helden* (1882).

Rainbows as beings

The Graeco-Roman goddess Iris' status as a personification of the rainbow is questionable, with some ancient sources citing only her use of the rainbow as a bridge, and others her wearing of the rainbow as a garment. A handsome prince, enchanted in such a way that he can only speak in the presence of a rainbow or spectrum, and who is therefore known as 'Prince Rainbow', occurs in one French fairy tale from before 1725.[22] But interestingly, other mythic occurrences of rainbows as humanoids – as distinct from their personal possessions or body parts – are rare almost to the point of non-existence. A key exception proving this rule is the Buddhist yogic tradition that certain individuals, such as the Great Guru Padmasambhava, achieved a 'Rainbow Body' by transforming the 'five major organs and systems of the human body ... [into] five chakras':

Achievement of the Mahasukha [Rainbow Body]
means [to be] transformed into the highest physical
form, the rainbow. . . . The highest wisdom is the Light.
Sinful persons have no light, just darkness . . . The form
and color of the rainbow reflected in the no-cloud blue
sky is the physical aspect and the light is the wisdom.
The physical flesh body is transformed into the form
and color of a Rainbow Body while its light is the
wisdom . . . Physically, because he is a rainbow, there
is no death.[23]

Mikalojus
Konstantinas
Ciurlionis, *Thor*,
1909, tempera.

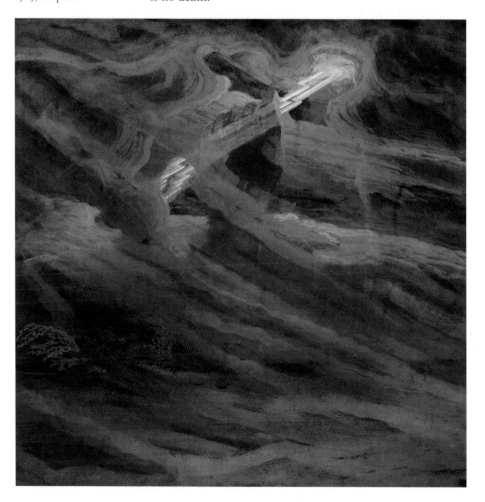

Pierre-Narcisse, Baron Guérin (1774–1833), *Morpheus and Iris*, 1811, oil on canvas.

It is worth noting in this context that the Buddha himself, or rather his aura, is associated with five colours: 'royal blue, golden yellow, dark red, white and magenta orange'.[24] Pre-Newtonian and non-European rainbows with as many as five colours (as in Chinese tradition) are rare, however, and mythic rainbows of six or more colours are as yet unrecorded.[25] Nevertheless, conceptions of the colours of the rainbow and their number are nearly as variable as culture itself. The ancient Greeks and others saw, or chose to see, a monochrome bow of red or purple. If there is a worldwide norm for mythic rainbows, it is probably for a three-coloured bow in which one of the colours is red, though two- and four-colour versions are also fairly common. For West African Yoruba people of the transatlantic diaspora, 'the image of the rainbow serpent Òṣùmàrè signifies the continuity of life and the blessings of the ancestors', and its four colours are ranked in terms of seniority: the oldest and therefore most important is white, followed by red, blue and yellow.[26] Those Yoruba who accept that black is a colour conceive of it as an extremely intense form of red or blue, and therefore it is ranked second or third after white.[27]

With regard to the 'continuity of life' more specifically, Òṣùmàrè – who is either a seasonal gender-shifter or hermaphrodite, according to different Yoruba sub-groups – is

Buddhist flags at Ghost Festival, Guangji Temple, Beijing.

Gajjimare dancers,
West Africa,
before 1915.

associated with human procreation and the umbilical cord.[28]
As rainbow-serpents of the world go, however, s/he is perhaps
exceptionally benign, appearing in insurance advertising in
Benin, while the popular Nigerian music group Osumare trans-
late their name merely as 'rainbow' – seemingly without any
intent to go beyond the general sense of positivity and inclusiveness that this word and image tend nowadays to arouse
among Europeans and North Americans.[29] This may have to
do with the fact that, in the twenty-first century, worship of
Òṣùmàrè is 'more common in the Americas than it is in West

Africa'.[30] The same deity is known as Dan Ayido Houedo among the Fon-speaking peoples of Benin and southwestern Nigeria.[31] Like his non-rainbow-affiliated python-god predecessor Dangbe, whom he absorbed during the eighteenth-century military conquest of the Kingdom of Savi by the Kingdom of Dahomey, Dan Ayido Houedo controls movement, both between the earthly plane and the afterlife, and between different states of wealth and social status.[32] Other groups in both East and West Africa harbour similar beliefs, with the Hausas' Gajjimare – another gender-shifter – being perhaps the most similar.[33] In the highlands of central Haiti, the sex-shifting African deity is normalized as a couple: Ayida Wedo is the female rainbow, and Damballah-Wedo her serpent-husband.[34] For the Mbuti of the Democratic Republic of Congo, meanwhile, the rainbow is unambiguously a 'terror-inspiring, man-murdering snake-monster that devours human beings and brings about catastrophes'.[35]

Among the Feranmin people of New Guinea, the rainbow-serpent Magalim is said to live in mountain pools; when roused, he causes earthquakes, rain and thunder, and rises from the waters to swallow the person who awakened him. Though the Earth's surface would collapse without him, he is the enemy of mankind (and especially women) and a principal cause of both madness and malaria. He is similar in shape to a python, but 'enormous', and all serpents, eels and rainbows are related to him. His scales, if captured, 'can give a warrior the shimmering quality of the rainbow so he cannot be targeted by arrows'. This set of beliefs, far from being unique to the Feranmin, is part of 'an old formulation which has circulated through the region for some time and now finds partial expression in various groups', not merely in other parts of New Guinea but throughout Australia, where it has been described as 'characteristic of Australian [Aboriginal] culture as a whole'.[36] One of the universal characteristics cited is the association of the creature with naturally occurring prismatic or iridescent materials such as quartz crystal and mother-of-pearl. The disease element (as distinct from violent assault) is less clear-cut, though Aboriginal groups around Perth attribute to the rainbow-serpent 'all sores and wounds for which they cannot otherwise account'.[37]

Certain rainbow-snakes of Indonesia, ravishers of women and bringers of insanity, guard stocks of gold under the earth, and can cause this gold to 'rise up in the form of a rainbow' from their mouths.[38] A similar mythos, including the pot of gold, was particularly widespread in northern British Malaya, where various peoples used at least eight different words for the rainbow, all of which also meant 'serpent'. As elsewhere in the region, these creatures were unlucky and caused fevers. Moreover, by the early 1960s, such beliefs had spread from the indigenous to the non-indigenous population.[39] In the Philippines, the Mangyan people of the island of Mindoro and the Tagbanua people of Palawan specifically associate rain during rainbow-snake appearances with the danger of disease, in a more clear and direct manner than, say, the Feranmin do.[40]

Uniquely, perhaps, the rainbow-serpent Wanamangura of the Talainji (Western Australia) is associated with moonbows as well as rainbows.[41] For the Lardil people of Mornington Island, the serpent is part of a cautionary tale about hospitality and the mistreatment of outsiders: because he refused succour to a foreign woman, Rainbow Serpent not only caused her death, but 'Rainbow Serpent's own Country ceased to exist after his fit of selfishness'.[42] Further exemplary tales of the creature – from Australia, Java, Bali and throughout the southwestern Pacific, as well as parts of China – are almost endless, while in India, 'whatever their other sectarian differences . . . Buddhists, Hindus, and Muslims share a common set of stories that joins earth mounds, serpents, rainbows, and treasure.'[43] Within the very broad region covered by the present paragraph, belief that the rainbow-serpent is essentially benign appears limited to Queensland.[44] That being said, even within Queensland, the Kabi regard the rainbow-serpent as 'tricky and malignant', a slaughterer of men and shatterer of mountains, who tends to be helpful only to those people who already possess magical powers of their own.[45]

The rainbow-serpent myths of the Americas appear to occupy a middle position between, on the one hand, those of

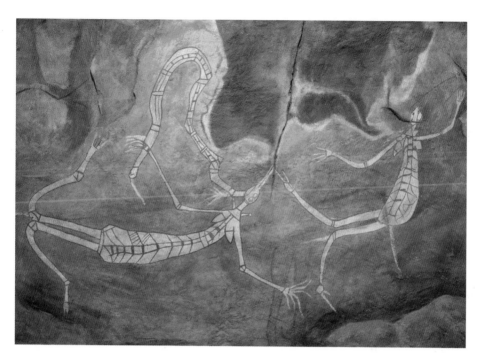

Replica of an Australian Aboriginal rock painting of Namaroto spirits and the rainbow-serpent Burlung, in the Anthropos Museum, Brno.

Africa (with a few key exceptions – for example, Mbuti, Zulu, Masai and Murle), in which rainbow-serpents tend to be benign; and on the other, those of the peoples of the southwestern Pacific Ocean and Southeast Asia, who overwhelmingly view such deities as both vicious and noxious.[46] Among South American groups, the Cumaná of Brazil and Guayakí of Paraguay seem to view their rainbow-serpents as especially dangerous, and the Panare of southern Venezuela use the same word to mean rainbow and 'were-anaconda'.[47] The Zuni mythology of the plumed water serpent, while it does not cite the rainbow specifically, is strikingly similar to various rainbow-serpent myths from the opposite side of the Pacific, previously referred to; and the Yaqui have stories of water-dwelling snakes with great spiritual power and 'rainbows on their foreheads'.[48] Many Amerindian groups including the Botocudo from eastern Brazil see rainbow-snakes as relatively unthreatening bringers of rain; but it would be simplistic to assume that a deity is benign simply because s/he is described as a 'bringer of rain', since such a deity if displeased

Rainbow in the
mountains of
Lachung, Sikkim,
May 1971.

might become rain-denying or even a bringer of drought.[49] Moreover, the desirability of rain may vary widely from place to place, from season to season, and from year to year, even within a particular belief-group.[50]

The passage of rainbow-serpent belief from West Africa to the slaveholding countries of the Western Hemisphere is readily explicable in historical terms. Other such trans-oceanic leaps are less so. One of the very few previous surveys of global rainbow mythologies managed to give an impression that the rainbow-as-malign-snake mythos had 'jumped' the South Pacific, some-how stretching from Australasia, Southeast Asia and Melanesia to Latin America without passing through northeastern Asia or the northern Americas en route – an apparent denial (albeit offhand) of the Bering Strait Land Bridge Theory of the peop-ling of the Americas.[51] A detailed analysis of roughly a thousand myths from the Bering Strait to Tierra del Fuego provides a different view entirely: in both North and South America, skunks and opossums are mythic opposites. North American myths associate the skunk with the burnt (and the rainbow) and the opossum with the rotten. Both have the power to resus-citate the dead. In South America, on the other hand, it is the opossum that is associated with the rainbow (to the point that in Guiana they share one name, *yawáre*), and both are credited with lethal power.[52]

John Loewenstein concluded that

> Myths of a giant rainbow-serpent . . . must be intimately connected with the occurrence and geographic distribution of a particular family of snakes, the Boidae . . . [This asso-ciation] may indeed have its origin in the brilliant colours and iridescent glow of most of these imposing reptiles, in their aquatic habits, their capacity to climb trees, and, last not least, in their habit to hibernate and reappear after the first seasonal rains.[53]

However, he goes on to admit that 'zoological evidence alone cannot explain' the mythos's 'almost world-wide distribution', and instead endorsed the idea that the rainbow-serpent myth could have had a 'very early prehistoric' origin, and been distributed worldwide with 'the first dispersal of mankind' out of Africa.[54]

University of Hawaii linguist Robert Blust specifically assigns a unitary, pan-human rainbow-serpent mythos to the Pleistocene epoch, and argues 'that dragons are the end point of a conceptual development which began with rainbows' and 'arose through processes of reasoning which do not differ essentially from those underlying modern scientific explanations.'[55] Blust asked,

> Why are dragons so often associated with waterfalls, pools, and caves? Why are they widely regarded as controllers of rain? . . . Why do they live in terrestrial water sources and yet take flight at the time of the rains? Why are they attacked by thunder or lightning? Why do they breathe fire? Why do they often guard a treasure, in particular a hoard of gold?[56]

Blust's arguments point up a major philosophical split within the study of myth, and in particular the study of geographically dispersed but very similar myths such as Rainbow-Snake and Rainbow-Bridge. Scholars – largely according to their own taste – have tended to ascribe the wide distribution of such stories to one of three processes: inter-cultural communication; simultaneous invention; or the fragmentation of a single myth that was once common to all people. Blust argues vigorously against the first possibility, on the grounds that 'global diffusion implies contacts for which we need some type of independent evidence', and in favour of a combination of the second and third elements, wherein the fragmentation of a single, universal mythic figure of perhaps 100,000 years ago – Rainbow-Serpent – led to the independent development of dozens of different dragon mythologies.[57] My own research, in contrast, indicates that the pre-modern diffusion of ideas and stories was constant, ubiquitous and mostly unremarked in official sources.[58] I would therefore suggest that the notion needing to be proved in the Rainbow-Serpent and Rainbow-Bridge scenarios is not the existence of long-range cultural diffusion, but rather the existence of long-term cultural isolation or impermeability. The rainbow-snake mythos was not entirely unknown even in Europe, with a variant being recorded in Estonia, while the Old English language had similar terms for the rainbow, *rén-boga*, and a coiled snake, *hring-boga*.[59] Even

Indian rainbow corn, Minnesota, 1937.

the giant Midgard Serpent of pre-Christian Scandinavia, famously defeated by the thunder-god Thor, can be argued to differ
from rainbow-serpents only in degree, on the grounds that
disparate world myths make rainbow-serpents the enemies or
opposites of thunder or lightning.[60] As such, it is intriguing to
compare the Graeco-Roman messenger-goddess Iris, who is
frequently identified with or as the rainbow, and the Norse
entities that seem to fulfil her function at various times and in
various ways: Yggdrasil the world-tree, Ratatoskr the sarcastic
messenger-squirrel and, of course, Bifrost itself.[61]

It would be a mistake to conclude this section without referring to the Rainbow Family of Living Light, a modern nomadic
tribe founded in or by 1972. Originally operating within the
United States, it had spread to Europe by 1983 and boasted

Rainbow Gathering
in Bosnia, 2007.

100,000 members worldwide in 1999.[62] Regular 'Rainbow Gatherings', whose attendees might or might not identify themselves individually as 'Rainbows', consist of

> a collection of camps including 'Jesus camps,' Krishnas, 'faeries,' camps of meat eaters and of vegetarians, substance-free camps, and camps of heavy drinkers. Yet the Gatherings as a whole are marked by a common commitment to radical equality and respect for diversity, personal freedom, and the environment.[63]

On the negative side, Rainbow life has been marked by 'perennial conflicts . . . with the u.s. Forest Service and the mass media' as well as an arguably racist 'appropriation and invention of American Indian traditions and iconography' (in Europe as well as North America), amid 'little willingness to deal respectfully with actual Indian communities'.[64]

Rainbows, luck and fear

The rainbow, perhaps due to the fact that it can still be seen everywhere, has escaped the fate of most European mythic figures in modern times – namely, to be 'excluded from everyday life and relegated to the vague and frightening zone where everything that was an object of religious belief was assumed to belong'.[65] Notions of the rainbow as unlucky prevailed well into the twentieth century in many parts of Europe, including England and Russia, especially among children.[66] Nor would it always be good luck to find the fabled pot of gold allegedly placed at the rainbow's end by an Irish leprechaun – a subtype of fairy that is itself 'a symbol . . . of luck' but which, like most other fairies, has a number of sinister characteristics.[67] Pointing at a rainbow was considered 'foolhardy' in Hungary, and unlucky or dangerous in Hawaii, Indonesia, China, Central America and Gabon, leading a recent scholarly work to characterize this act as a 'near-universal taboo'.[68] And in the relatively recent folklore of Albania, Bohemia, France, Hungary and Serbia, rainbows

Rainbow and leprechaun parade float in France, 2015.

were believed to cause unwelcome, or at any rate unexpected, sex changes in humans.[69]

The evidence for pre- or non-Christian survivals, within Europe, of a menacing folkloric rainbow could probably be multiplied. In any case, the bow of God's covenant is perhaps menacing enough in its implications. For instance, the covenant does *not* prohibit God from destroying life on Earth via plague, earthquake or indeed anything other than flood, as the title of James Baldwin's book *The Fire Next Time* (1963) pointedly reminds us. It was claimed in the early Middle Ages that no rainbow would be seen in the forty years immediately preceding

Albrecht Dürer (1471–1528), *Melencolia I*, 1514, engraving. As well as the rainbow, the winged woman representing melancholy is accompanied by a ladder, hourglass, bell, magic square, carpenter's tools, a sleeping dog, and a putto with a notebook and inkwell.

the Apocalypse, and also that the predominately red colour of the bow indicated that the world would be ended by fire.[70] As such, it is worth wondering whether the two dominant modes of engagement with the rainbow today – as a harmless and (mostly) explicable scientific phenomenon, and as cute, cuddly kitsch – both equally serve to paper over the awe and fear that rainbows have inspired throughout human history and in every part of the world.

4 Rainbows in Literature, Poetry and Music

Can all that optics teach, unfold
Thy form to please me so,
As when I dreamt of gems and gold
Hid in thy radiant bow?

When Science from Creation's face
Enchantment's veil withdraws,
What lovely visions yield their place
To cold material laws!
Thomas Campbell (1777–1844)[1]

Literary references to the rainbow are as old as literature itself. First written down in the eighteenth or nineteenth century BCE, but with some content probably nine centuries older still, *The Epic of Gilgamesh* is an adventure story with strong supernatural elements. It prefigures the later Greek tales of Prometheus, Hercules and Odysseus, as well as – in the character of Utnapishtim – the biblical Noah. 'I slaughtered a sheep, and roasted it for the gods,' Utnapishtim tells the titular hero regarding the aftermath of a great world-destroying flood.

So pleased were they, that down came fertile Eanna
With necklace of lapis lazuli, gold, and amythest, which
Thereinafter we called the rainbow
As this artisan gift she gave to us.
A gift she gave to ever remind us of dimmer sadder days.[2]

Eanna, also known by the name Inanna, had been the most important female deity in Mesopotamia since at least the fourth millennium BCE.

We have already discussed the goddess Iris in the classical tradition, but the Renaissance saw a revival of interest in her, alongside other aspects of Graeco-Roman mythology and

Eleventh tablet of the *Epic of Gilgamesh*, containing the story of Utnapishtim and the Flood.

civilization far too numerous to list. *Virgilius*, an anonymous English prose romance of 1518, features a 'brygge in ye ayer' (bridge in the air) – identified by one critic as 'reminiscent of the rainbow of Iris and such rainbow-bridges as Bifrost' – which is used by a sultan's daughter to travel from Babylon to Rome. This is unusual, however: most similar tales of the early modern era replace the bridge with a magic carpet or hat.[3]

A much more significant appearance of the ancient rainbow-goddess was in William Shakespeare's *The Tempest*, first performed in 1611. According to Homer's contemporary Hesiod, who created the first systematic account of Greek mythology around 700 BCE, Iris is a sister of the vile harpies; and this sinister background may relate directly to the ominous potential of the supposedly 'white' magic used by Prospero, *The Tempest*'s central figure and Iris' human protégé.[4] In Act III, Scene 3,

Prospero's familiar, Ariel – hitherto invisible – enters 'like a harpy' amid thunder and lightning, and is referred to by his master as performing the 'figure of' a harpy. Iris herself appears in Act IV, Scene 1, acting as a herald to the more senior goddesses Hera and Ceres. Shakespeare's Iris describes herself as a 'watery arch', and is described by Ceres as 'many-colour'd', though the only colours mentioned by name are 'saffron' and blue: ascribed by Ceres to Iris' wings and bow, respectively. Even within the play, however, the accuracy of these elements vis-à-vis ancient mythology is called into question, with Prospero referring to Iris and Ceres as 'Spirits, which by mine art / I have from their confines call'd to enact / My present fancies' (Act IV, Scene 1), rather than actual goddesses. The titular storm is quickly revealed as an illusion as well, raising questions of whether the play as a whole is a reference, perhaps parodic, to the Bible's story of Noah. Act IV's wedding sequence presents

A representation of the Dharmachakra, with a symbol of the goddess Ishtar at the centre.

the meeting of earth and heaven under the rainbow,
the symbol of Noah's new-washed world, after the
tempest and flood had receded, and when it was
promised that springtime and harvest would not
cease . . . Out of the cycle of time in ordinary nature
we have reached a paradise . . . where spring and autumn
exist together.[5]

Not popular in its own time, *The Tempest* became well known
(along with expanded versions and parodies of it) only during
Britain's 'long eighteenth century' that began with the restoration
of the monarchy in 1660. This may explain the somewhat long
jump taken by rainbows within English literature, from Shakespeare and his near-contemporary William Drummond to the
Romantic poets of the late Georgian era – upon whom the
influence of the play seems to have been fairly direct.[6] Even John
Milton's contribution seems more a description of Shakespeare's
Tempest characters than an original poetic insight:

I took it for a faery vision
Of some gay creatures of the element,
That in the colours of the rainbow live,
And play i' th' plighted clouds.[7]

And in the whole century and a half after the end of the English
Civil War, only two really popular English-language ballads
contained the term 'Rain[e]bow' in their titles. One was about
the real-life rape and murder of a woman who happened to be
named Elizabeth Rainbow; the other celebrated a battle involving an English vessel called the *Rainbow*, one of at least eleven
British and Commonwealth warships that have been named
after the phenomenon to date.

None of this is to suggest, however, that rainbow poetry
ceased to be written on straightforwardly biblical lines in
the early modern period or at any time afterwards. Thus, for
example, *The Rainbow* by Henry Vaughan (1622–1695) states,

Benjamin Smith's
1797 engraving of
Act I, Scene I of
William Shakespeare's
The Tempest, after
a painting by
George Romney.

When thou dost shine, darkness looks white and fair,
Storms turn to music, clouds to smiles and air

. . .

Bright pledge of peace and sunshine! the sure tie
Of thy Lord's hand, the object of His eye!
When I behold thee, though my light be dim,
Distant, and low, I can in thine see Him,
Who looks upon thee from His glorious throne,
And minds the covenant 'twixt all and One.[8]

Indeed, as George Landow has suggested, Vaughan's poem
formed an integral part of 'a revered tradition of devotional
poetry' which reminded its readers and hearers that the rainbow
'served not only to record God's covenant with Noah but also . . .
the second, or new, covenant, brought by Christ'. In England,
'this tradition seems to have retained vitality longer than any-
where else,' regardless of whether the audience was High Church,
Low Church or Roman Catholic.[9]

The 'million-coloured bow' and English Romantic poetry

One of Percy Bysshe Shelley's final poems, 'To a Lady, with a Guitar' (1822), takes the form of a speech made to Miranda, Prospero's daughter, by Ariel. Though 'To a Lady' does not actually describe a rainbow, it seems constantly to be on the verge of doing so, with its images of April showers, 'pattering rain, and breathing dew'.[10] Shelley's 'The Cloud' (1819) was more explicit:

> From cape to cape, with a bridge-like shape,
> Over a torrent sea,
> Sunbeam-proof, I hang like a roof,
> The mountains its columns be.
> The triumphal arch through which I march
> With hurricane, fire, and snow,
> When the Powers of the air are chained to my chair,
> Is the million-coloured bow;
> The sphere-fire above its soft colours wove,
> While the moist Earth was laughing below.[11]

Shelley's very first poetic publication, the banned *Queen Mab* (1813), presented a very different take on the rainbow – imagining it as emblematic of the religious fanaticism of a bygone and detestable age:

> And frantic priests waved the ill-omened cross
> O'er the unhappy earth: then shone the sun
> On showers of gore from the upflashing steel
>
> . . .
>
> And blood-red rainbows canopied the land.[12]

Shelley's friend George Gordon, Lord Byron (1788–1824), was apparently first inspired to write poetry in July 1799, at the age of eleven, upon meeting his beautiful thirteen-year-old cousin Margaret Parker: 'she looked,' he later wrote, 'as if she had been made out of a rainbow.'[13] The adult Byron returned to this image of incestuous-beloved-as-rainbow in *The Bride of*

Abydos (1813), with the heroine Zuleika upgraded from a cousin to a sister:

> Thou, my Zuleika! share and bless my bark;
> The Dove of peace and promise to mine ark!
> Or, since that hope denied in worlds of strife,
> Be thou the rainbow to the storms of life!
> The evening beam that smiles the cloud away,
> And tints to-morrow with prophetic ray![14]

However, it was their older fellow Romantic, William Words-worth (1770–1850), who wrote the most enduringly famous rainbow poem of the era, 'My Heart Leaps Up' (1802):

> My heart leaps up when I behold
> A rainbow in the sky:
> So was it when my life began;
> So is it now I am a man;
> So be it when I shall grow old,
> Or let me die!
> The Child is father of the Man;
> And I could wish my days to be
> Bound each to each by natural piety.[15]

This sense that there would always be more rainbows, however, was neatly balanced by the fact that they would appear only occasionally, at and for random intervals. As Wordsworth him-self put it, in his 'Ode on Intimations of Immortality from Recollections of Early Childhood', 'The rainbow comes and goes, / And lovely is the rose.'[16]

Samuel Taylor Coleridge (1772–1834), who penned 'Hymn Before Sunrise in the Vale of Chamouni' in the same year that Wordsworth wrote 'My Heart Leaps Up', contrasted the beauty of his setting's rainbows (and occasional flowers, goats, eagles and lightning) against the 'dread and silent . . . dark and icy' black-and-white Alpine landscape, with its '[m]otionless cataracts', 'silent torrents' and 'precipitous, black, jagged rocks,

/ For ever shattered and the same for ever'. But where the atheist Shelley saw the wonders of the sky and weather as self-explanatory, or at least requiring no supernatural explanation, Coleridge ascribed them firmly and repeatedly to 'God!' He described Mont Blanc as the 'dread ambassador from Earth to Heaven, / Great Hierarch',[17] in perhaps a further nod to the biblical rainbow covenant and the balanced relations between Man and God that upward-pointing bows, and exchanges of ambassadors, tend to suggest. Coleridge's sense of the bow as a Heaven/Earth interchange, underwritten by divine grace, was strongly reinforced a quarter-century later in his 'The Two Founts':

> As on the driving cloud the shiny bow,
> That gracious thing made up of tears and light,
> Mid the wild rack and rain that slants below
> Stands smiling forth, unmov'd and freshly bright;
> As though the spirits of all
>   lovely flowers,
> Inweaving each its wreath and
>   dewy crown,
> Or ere they sank to earth in
>   vernal showers,
> Had built a bridge to tempt
>   he angels down.[18]

An 1857 depiction of Lord Byron's characters Selim and Zuleika from *The Bride of Abydos*, painted by Eugène Delacroix (1798–1863).

In practice, of course, there is little to choose from between Coleridge's 'indifferent, absentee, Epicurean God' and Shelley's non-God; and Coleridge – despite being the son of a clergyman – explicitly rejected the still-popular view that rainbows were the result of 'an actual miracle'.[19]

Regardless of the direction of their thoughts on the nature or existence of the divine, however,

The Mer de Glace and the Valley of Chamonix as seen from the Chapieux, late 19th century.

the Romantics – not only in the literary sphere, or even in the English-speaking world – were stung by the relatively recent scientific explication of the rainbow as a natural phenomenon. John Keats (1795–1821) most famously expressed this anger and regret in *Lamia* (1820), an extended allegory of science's destruction of beauty and happiness:

> ... Do not all charms fly
> At the mere touch of cold philosophy?
> There was an awful rainbow once in heaven:
> We know her woof, her texture; she is given
> In the dull catalogue of common things.

Philosophy will clip an Angel's wings,
Conquer all mysteries by rule and line,
Empty the haunted air, and gnomèd mine –
Unweave a rainbow, as it erewhile made
The tender-person'd Lamia melt into a shade.[20]

Rainbows in early American letters

Whether or not one accepts that North America had its own Romantic movement, Keats's *Lamia* strongly influenced the young Anglo-American author Edgar Allan Poe (1809–1849), and in particular his sonnet 'To Science' (1829): 'Why preyest thou thus upon the poet's heart, / Vulture, whose wings are dull realities?'[21] Mark Twain (1835–1910), following in Coleridge's physical footsteps around Chamonix and Mont Blanc, echoed similar sentiments in prose:

We have not the reverent feeling for the rainbow that a savage has, because we know how it is made. We have lost as much as we gained by prying into that matter.[22]

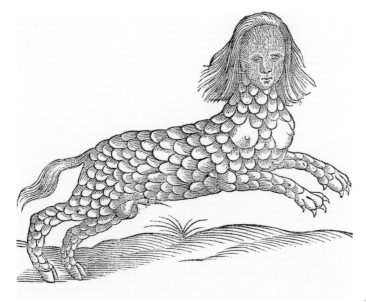

Lamia, from Edward Topsell's 17th-century *The History of Four-footed Beasts.*

This sense of regret carries even greater weight in light of Twain's well-documented love of science, his pioneering role in the genre of science fiction, and his close friendship with Nikola Tesla. The American proto-environmentalist author Henry David Thoreau (1817–1862), meanwhile, claimed to have 'clutched' the rainbow, as part of the 'intangible and indescribable' but 'true harvest' of his daily life.[23]

However, nineteenth-century America's most interesting literary use of the rainbow – or perhaps of colour in general – was by Nathaniel Hawthorne (1804–1864).[24] A New Englander embarrassed by the pugnacious intolerance of the region's Puritan heritage, including especially the actions of his own ancestor who had been a judge at the notorious Salem Witch Trials, Hawthorne wrote 'The May-pole of Merry Mount' in 1837.[25] The short story opens with non-Puritan Anglican colonists in the vicinity of modern-day Quincy, Massachusetts, celebrating a wedding in the forest, at a rather extraordinary may-pole: it is a tall and slender living pine tree, decorated with 'a silken banner colored like the rainbow', 'ribbons . . . in fantastic knots of twenty different colors, but no sad ones', and 'stained with the seven brilliant hues of the banner at its top'. The groom wears 'a scarf of the rainbow pattern', and the bride's garb, though not so clearly described, is said to be similar. Their wedding guests are disguised in masks of forest animals or grotesque human faces. Though the party intend nothing more than the maintenance of the traditions of their homeland – underlined by the presence of 'roses reared from English seeds' and an 'English priest' – they are perceived as 'devils' by a war-band of Puritans watching them from the shadows. And as night draws on, 'some of these black shadows . . . rushed forth in human shape.'

> The future complexion of New England was involved in this important quarrel. Should the grisly saints establish their jurisdiction over the gay sinners, then would their spirits darken all the clime and make it a land of clouded visages, of hard toil, of sermon and psalm for ever; but should the banner-staff of Merry Mount be fortunate,

sunshine would break upon the hills, and flowers would beautify the forest and late posterity do homage to the Maypole.

But the rainbow is not yet to have its victory. The Puritan para-militaries surprise and swiftly overwhelm the revellers, shooting their pet bear through the head, forcibly cutting the bride-groom's hair 'in the true pumpkin-shell fashion', and hacking the maypole down with a sword. Finally, the roses are tossed aside by an iron-gauntleted hand, in what Hawthorne describes as 'a deed of prophecy': presumably, of the Cromwellian-tinged American Revolution to come, and its legacy of bitter tension between collective security and individual freedom that has never been adequately resolved. Interestingly, Hawthorne does not side unambiguously with either group, and seems to condemn both at different times and in different ways. However,

Maypole dancers, Llanfyllin, 1941.

the proto-hippies seem on balance to have his greatest sympathy. Resonating more in the twentieth century than its own, 'The May-pole of Merry Mount' was adapted as an opera by Howard Hanson in 1933, and as a play by Robert Lowell in 1964.

Hawthorne's masterpiece, *The Scarlet Letter* (1850), describes its adulterous seventeenth-century heroine's passion as having acted as a prism,

> the medium through which were transmitted to the unborn infant the rays of its moral life; and however white and clear originally, they had taken the deep stains of crimson and gold, the fiery lustre, the black shadow, and all the untempered light of the intervening substance . . . even some of the very cloud shapes of gloom and despondency that had brooded in her heart.[26]

Another ambiguous rainbow analogy occurs in Hawthorne's *House of the Seven Gables* (1851). The novel's central character Hepzibah Pyncheon, a gentlewoman who lives in poverty with her brother – a convicted murderer who has served out his thirty-year sentence – finds herself continually torn between tenderness and coldness, sorrow and joy; but sometimes, 'the laughter and tears came both at once, and surrounded our poor Hepzibah, in a moral sense, with a kind of pale, dim rainbow.'[27]

Rainbow poetry after the Romantics

A completely circular spray-rainbow glimpsed in boyhood while looking down into the Zambezi River from a height became the subject of F. C. Kolbe's religious poem 'A Rainbow at Victoria Falls' (1906); and a moonbow observed from a speeding train in April forms the central motif of one of the great works of modern Spanish poetry, 'Iris de la noche' by Antonio Machado (1875–1939). In the latter, the poet seems to suggest that humanity's

> present weakness . . . lies in the diffuseness of our light. The source of our future strength lies in the congregation

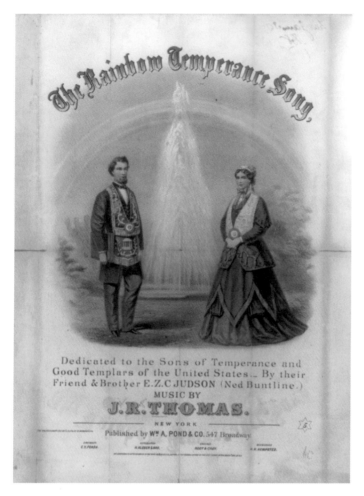

A latter-day puritan rainbow. Lithograph by Major & Knapp, 1868.

of every individual gleam, all bound together in one all-encompassing band. Machado's God has a form – a unifying circle of intelligence whose symbolic representation is a ring around the moon.[28]

Similarly, in Robert Browning's 'Christmas-eve and Easter-day' (1850), the poet – fresh off a train from Manchester with its 'thump-thump and shriek-shriek' – enters a Nonconformist chapel to avoid a rainstorm. On fleeing back into the storm to escape the glares of the filthy and censorious congregation and 'the pig-of-lead-like pressure / Of the preaching-man's immense stupidity', Browning sees a 'moon-rainbow vast and perfect'. As if this apparition itself were not enough to symbolize a faith that transcends mere religious observance, he adds,

> All at once I looked up with terror.
> He was there.
> He himself with his human air.
> On the narrow pathway, just before.
> I saw the back of him, no more –
> He had left the chapel, then, as I.
> I forgot all about the sky.
> No face: only the sight
> Of a sweepy garment, vast and white.[29]

Amid the decline of faith, or at least of faith-related poetry, in the post-war Western world, uses of the rainbow as a poetic image have been diffuse and deeply personal. Antonio Skarmeta's *Los días del arcoiris* (The Days of The Rainbow, 2011) deploys it as a symbol of hope that the end of the dictatorship of General Augusto Pinochet of Chile will bring a better future, and in Mario Benedetti's poem *Arco iris*, it is used to describe the smile of a lover who has been crying. Ogden Nash's characteristically funny 'Song to Be Sung by the Father of Infant Female Children' begins with a parodic reference to Wordsworth's 'My Heart Leaps Up'; and in 'The Watergaw', Hugh MacDiarmid addresses a dead person whose 'last wild look' was recalled to mind by the 'foolish' light of the broken or partial rainbow of the title.

The Pulitzer Prize-winning American poet Carl Sandburg, in what may have been a consciously Wordsworthian yearning for rainbows to stay put, said, 'Poetry is a phantom script telling how rainbows are made and why they go away.'[30] A posthumous volume of his work was given the title *Rainbows Are Made* in 1982; the degree to which this twists the meaning of Sandburg's original comment is more or less in the eye of the beholder.

Rainbow at Victoria
Falls, 2012.

Notable literary rainbows of the twentieth century

In his acclaimed novel *Howard's End* (1910), E. M. Forster wrote of 'the building of the rainbow bridge that should connect the prose in us with the passion.' Without this bridge, he went on,

> we are meaningless fragments, half monks, half beasts, unconnected arches that have never joined into a man. With it love is born, and alights on the highest curve, glowing against the grey, sober against the fire. Happy the man who sees from either aspect the glory of these outspread wings.[31]

But arguably the most significant literary use of the rainbow in the early twentieth century was by D. H. Lawrence (1885–1930) in his novel *The Rainbow* (1915), a multigenerational family saga depicting the fall and rise and fall of a family of semi-aristocratic Polish refugees who intermarry with yeomen farmers in the North Midlands of England in Victorian times. Perhaps following Forster's lead, the book deploys a complex pattern of arch metaphors in support of its plea for unmediated human feeling, especially in opposition to what the female characters perceive as the 'heap of dead matter' that is organized religion: with the 'broken' arches of Gothic churches and the binding, excluding 'closed circle' of the wedding ring coming in for special criticism. Against these representatives of coldness and deadness, Lawrence sets the rainbow, and a desirable young woman's eyebrows, and even – paradoxically – the Romanesque arches of older, presumably Catholic churches. This last juxtaposition is especially interesting, since Lawrence's Catholic émigrés convert to Protestantism on arrival, without hesitation and often fervently, one of them even becoming a vicar. Rainbows are repeatedly and explicitly compared to architectural arches and doorways, both real and imaginary:

> Anna loved the child very much, oh, very much. Yet still she was not quite fulfilled. She had a slight expectant feeling, as of a door half opened. Here she was, safe and

still in Cossethay. But she felt as if she were not in Cossethay
at all. She was straining her eyes to something beyond . . .
A faint, gleaming horizon, a long way off, and a rainbow like
an archway, a shadow-door with faintly coloured coping
above it. Must she be moving thither?[32]

On hearing the biblical rainbow story, one of *The Rainbow's*
principal female characters, Ursula Brangwen, silently derides
Noah and his sons Shem, Ham and Japheth for their greed to 'be
masters of every thing, sub-tenants under the great Proprietor'.
Later, she condemns neo-Gothic church and school architecture
as a pure expression of 'domineering . . . vulgar authority'. Christ
is explicitly compared to the Moon, not to the rainbow; but
notably, He is a crescent moon in particular. The book concludes
with Ursula, annealed by the love and loss of a rather blank young
army officer, beholding a 'faint, vast rainbow' form over 'the
corruption of new houses' and the 'hideous obsoleteness' of the
old church tower in her grim colliery town. A practised hater of
religion, she nevertheless responds to this vision with what
might be called a secular variant of apocalyptic ecstasy:

> And the rainbow stood on the earth. She knew that the
> sordid people who crept hard-scaled and separate on the
> face of the world's corruption were living still, that the
> rainbow was arched in their blood and would quiver to
> life in their spirit, that they would cast off their horny
> covering of disintegration, that new, clean, naked bodies
> would issue to a new germination, to a new growth, rising
> to the light and the wind and the clean rain of heaven.
> She saw in the rainbow the earth's new architecture, the
> old, brittle corruption of houses and factories swept away,
> the world built up in a living fabric of Truth, fitting to the
> over-arching heaven.[33]

It would be difficult indeed to devise a better bridge linking the
'First New Age' of the Edwardian era both backwards in time,
to the political radicalism that attended early Protestantism, and

forwards, not merely to the New Age or Age of Aquarius of the 1970s and '80s, but to the horrors of the twentieth-century battlefield.

For Texas Chicano author Genaro González, whose acclaimed novel *Rainbow's End* of 1988 is a multigenerational saga beginning with heroic illegal immigration through floodwaters during the Great Depression, the titular 'end' seems to refer to the present-day 'acceptance of mall culture and assimilation which signifies the rejection of the Mexican heritage', an interesting sidelight on the positive rainbow imagery with which multiculturalist or 'melting-pot' rhetoric is so often associated in the political sphere.[34] From at least 1959, which saw the publication of Alan Lomax's *The Rainbow Sign* – a magisterial oral history of African American folklore – 'rainbow' has been among the 'go-to' title words for printed works on multiculturalism or minority cultures, a typical example being Kenneth Rosen's *Voices of the Rainbow: Contemporary Poetry by American Indians* (1975). A prominent 'black-oriented' cultural centre called Rainbow Sign, possibly after Lomax's book, was founded in an old mortuary on Grove Street in Berkeley, California, in the early 1970s; it hosted 'book parties', receptions for authors including James Baldwin, jazz bands and at least one 'seventy-two-hour poetry marathon', and was remembered with immense fondness in the greyer decade that followed.[35] Also in northern California, in 1974, Ntozake Shange wrote a cycle of poems for live performance, collectively entitled *For Colored Girls Who Have Considered Suicide/When the Rainbow is Enuf*. It featured seven nameless black female characters identified only by their costume colours: red, orange, yellow, green, blue, purple and brown. In 1976 *For Colored Girls* . . . became the second show written by an African American woman ever to open on Broadway, and the first since 1960.[36] Its title was inspired by an actual double rainbow that Shange saw while driving on California's famed Highway One, causing her to be 'overcome' by 'a feeling of near death or near catastrophe'.[37] Others, however, were quick to stress a less mythic viewpoint. Addressing Shange directly, one enthusiastic reviewer of the book version of 1977 commented, 'I, too, have considered suicide, but . . . I found, glistening at [the rainbow's] end, not a pot of gold, but *me!*'[38] By now a Tony Award-winning play, the work was later adapted for television and film, with mixed success.

Direct personification of the rainbow remains as rare in Western culture today as it was in medieval times. However, exceptions exist to every rule. A fallen rainbow named Siafra, laid low by ozone depletion above our 'fading globe of smog' hemmed in by radio waves, lands with a bang on her 'large, lovely and exceeding colourful rump' in Jonathan Enright's amusing yet downbeat short story 'The Rainbow' (2000). On meeting a leprechaun, the music-loving rainbow asks him for some gold, arguing that his thrift was useless 'as all the world was dying'. Agreeing with Siafra's defeatist logic, the leprechaun flees into another dimension, but not before giving her some gold, which she spends on a ticket to a Courtney Love concert in Seattle.[39]

Rainbows in children's literature

Like one's own reviews, one should not judge rainbows in children's books by whether they are good or bad, but by their weight. Having originally intended to provide individual analyses of the deployment of the rainbow in illustrated books for children, I soon realized that, among those published since the 1990s, rainbow-less works appear to be in the minority. Even the tongue-in-cheek masterpiece *The Day the Crayons Quit* (2014) by Drew Daywalt and Oliver Jeffers features a pair of all-black rainbows. It was not always thus; I cannot specifically remember any rainbows in the favourite books of my own childhood in the 1970s, most of which had been published to popular acclaim in the 1940s, '50s and '60s. Two major English children's stories of the pre-First World War period featured the rainbow as a central motif, however, and it would be hard to say which is the creepier.

George MacDonald's 'The Golden Key', a short story published in 1867, is set in Fairyland. The titular key appears at the rainbow's end (which in Fairyland can be readily approached),

Now largely forgotten, the high-imperialist *Where the Rainbow Ends* was one of the most popular British children's stories in the run-up to the First World War.

and no matter how often this key is taken away, it appears again. 'Perhaps it is the rainbow's egg,' speculates Mossy, the main male character.[40] The rainbow in Fairyland has the usual seven colours, plus 'shade after shade beyond violet', and above the red, something 'gorgeous and mysterious . . . a colour [Mossy] had never seen before'. Moreover, when the Sun set, 'the rainbow only glowed the brighter. For the rainbow of Fairyland is not dependent upon the sun, as ours is.' And inside it, 'beautiful forms' of men, women and children can be seen 'slowly ascending as if by the steps of a winding stair'. Mossy and his ten-year-old female counterpart, Tangle, embark on a series of adventures in search of the 'the country whence the shadows came', many involving rainbow-feathered 'air-fish', whose ambition – in common with the other animals of Fairyland – 'is to be eaten by the people'. As often happens in fairy stories, the boy and girl grow old in the space of hours, but having thus 'tasted of death', Mossy proclaims it 'better than life'. From this point, he acquires the ability to walk on water. With a kind of sickening inevitability, our heroes find the lock that fits their golden key, and open it.

> They were in the rainbow. Far abroad, over ocean and land, they could see through its transparent walls the earth beneath their feet. Stairs beside stairs wound up together, and beautiful beings of all ages climbed along with them. They knew that they were going up to the country whence the shadows fall. And by this time I think they must have got there.[41]

Far from blunting the sinister aspect of the fairy death-rainbow, the bloodthirsty imperialism of *Where the Rainbow Ends* – a children's play and book of 1911 by Clifford Mills and John Ramsey – merely adds another layer to it, endorsing British participation in an arms race against Germany, France, the USA and the Russian Empire. The 'small and stunted' but 'British-born' lion belonging to the child heroes (one of whom is a naval cadet) is used as a none-too-subtle metaphor for this.[42] Those 'whose heart is pure and faith strong' will find the titular land, a barely veiled version of the afterlife, where 'all lost loved ones are found'.[43] After a journey by magic carpet, accompanied by Saint George, they must face a hotchpotch of supernatural beings including Sea-witch, Lake King, the will o' the wisp, gnomes, three types of elf, dragons, dragon-men, an enchanted tree that was previously a dragon, a genie, giant spiders and slimy worm-men, in addition to merely dangerous real-world animals such as panthers and hyenas. Finally,

> Reassured and upheld by faith in their champion, the awe-struck children stood, not daring even to glance at

the yawning chasm beneath them, their gaze following St George's upraised sword; and as they so stood, night with its terrors forsook the sky and across the tender, dawn-lit morn a glorious rainbow slowly shaped and grew till the thousand summits of the cragged mountains shone all roseate in its radiant glory. . . . For this was the children's shore; where in anticipation that knows not the fret of hope delayed[,] they await the coming of those who have loved and lost them.[44]

The play version went on to be professionally staged every Christmas season for 47 of the next 49 years. A silent film appeared in 1921, and at a stage performance attended by Queen Mary and Princess Elizabeth in 1937, a special rainbow-citing verse was added to 'God Save the King'.[45]

Special mention should also be made of *Very Hungry Caterpillar* author Eric Carle's *I See a Song* (1973), which explicitly connects rainbows to music. This was also released as a short animated film under the same title in 1993, set to an original classical composition by Julian Nott.

Rainbows and music

Given that Kepler and Newton (among others) both derived insights about the rainbow and its colours from their understanding of music, it is somewhat surprising that the reverse of this – the overt use of rainbow physics to inform musical composition – has been so rare. One important exception is Jacob ter Veldhuis's *Rainbow Concerto for Cello and Orchestra*, which premiered in Rotterdam in February 2003, conducted by Alexander Lazarev.

Ter Veldhuis treats the quiet solo cello opening of the Rainbow Concerto like a single color from which the entire visual spectrum emanates. The second movement, played *attacca,* gradually builds in intensity, rhythmic energy, and an ever-expanding orchestral palette. Musical cells repeat, grow and evolve like shimmering particles of light reflecting the larger whole. The result is a concerto requiring tremendous musical concentration and sustaining power more than sheer manual virtuosity.[46]

Hector Berlioz (1803–1869) associated colours with musical timbre, and once argued that 'instrumentation is to music precisely what colour is to painting.'[47] Claude Debussy (1862–1918) has also been identified as highly colour-aware.[48] But this did not apparently result in compositions directly inspired by rainbows.

In the twentieth century and since, popular music has arguably been the Western world's main vehicle for the production and reception of poetry. As with poetry per se, rainbows feature in it in myriad ways, some of them unexpected. In 'She's a Rainbow' (1967), the Rolling Stones made a straightforwardly Byronic association between the rainbow and a female beloved; and perhaps following Shelley's *Queen Mab*, or some even older tradition, Neil Young's 'Down by the River' (1969) used the bow as an emblem of emotional turmoil that ends in murder. Indeed, there were only a handful of really famous songs of the 1960s – notably 'Lazy Sunday Afternoon' by the Small Faces – that deployed rainbow-citing lyrics in the fuzzily positive vein that would soon come to afflict popular culture more generally. In 1970, as the Age of Aquarius dawned, rainbow-climbing would stand in for rainbow-chasing in Bread's number-one hit single 'Make It with You'. The end of the rainbow appeared as a traditional symbol of hope in Badfinger's 'Carry on till Tomorrow' in the same year, and our tendency to regard the lyric in question as dark and ambiguous may be conditioned by the fact that two members of the band later committed suicide by hanging in separate incidents. But it would be Richard and Linda Thompson who used the rainbow to most devastating effect as a symbol of Baader-Meinhof-era malaise, in 'The End of the Rainbow' (1974). Addressed to an infant, the song maintains that in a civilization universally blighted by greed, corruption, loathing and interpersonal violence, there is not merely no pot of gold at the end of the rainbow, but nothing whatsoever. The gold-less world seamlessly becomes a goal-less one, in which there is no longer any reason for the child to become an adult, except perhaps to prove its sheer will to endure despite a lack of familial love that borders on outright aggression. Indeed, the family atmosphere that the song describes leads the listener to question whether the singer is actually benevolent, or yet another horrible relative, who in lieu of good advice offers pure negativity, wrapped up in a bow.

Even more surprising, perhaps, is the degree to which rainbows have featured in heavy metal music, a rock subgenre better known for monochromatic black leather. The metal band Rainbow, formed in Hertfordshire in 1975 by Deep Purple alumnus Ritchie Blackmore, had a sound that has been called 'medieval' or 'gothic', with a lyrical emphasis on 'demons and wizards'.[49] Though urban legend assigns the naming of the band to Hollywood's Rainbow Bar & Grill, their classical-music influences may suggest a deeper connection to the rainbow phenomenon; this is perhaps corroborated by their use – beginning on the first date of their first world tour – of a 12 m (40 ft) rainbow-shaped stage prop containing 3,000 computer-controlled light bulbs. Ex-Rainbow members Ronny James Dio and Jimmy Bain went on to score a major hit with the single 'Rainbow in the Dark' in 1983, and their concert film *A Special from the Spectrum* of 1984 sold more than 50,000 VHS copies. Other relatively recent recording artists to gain major airplay with rainbow-citing songs have included Peggy Lee, Cyndi Lauper, Paul de Leeuw and Mariah Carey. As a pop-cultural phenomenon that transcends music per se, Yip Harburg's 'Somewhere over the Rainbow' will be dealt with in the 'Rainbows in Film' section of Chapter Five.

An interesting early rainbow pop song is Cole Porter's song 'Down in the Depths on the Ninetieth Floor'. Part of the musical *Red, Hot and Blue*, which was performed more than 180 times on Broadway in 1936–7 and starred Ethel Merman, Bob Hope and Jimmy Durante, the song contained an early warning of the tidal wave of kitschy neon advertising that would nearly overwhelm the rainbow as a symbol in the decades that followed. Such critiques have been fairly rare, however; and in the visual sphere, at least, musicians are more likely to participate in than to oppose the use of the rainbow as an eye-catching, vaguely positive expression of nothing much in particular.

One noteworthy exception was the Ken Kelly cover artwork that the band Rainbow chose for their second album, *Rising* (1976), on which a manifestly tangible five-coloured bow is seized in the deathly grip of an immense hand covered in long, dank fur, bursting upward from the sea. Another was the seven-banded rainbow executed entirely in four shades of grey that appeared in Robert del Naja's cover art for Massive Attack's *Heligoland* album of 2010, above a disturbing image of a man's face, one eye apparently weeping blood and the other circumscribed by a cog – the latter, perhaps, a reference to the mascara worn by the main character in Stanley Kubrick's long-banned film adaptation of Anthony Burgess's novel *A Clockwork Orange*. Acts as diverse as Jimmy Cliff, the Mahavishnu Orchestra

The iconic album cover of Pink Floyd's *Dark Side of the Moon* (1973).

and The Hand of Doctor Foxmeat Paints a Monochromatic Rainbow have also featured unusual bows in their cover art; and in an interesting closing of the circle, the Australian state of Victoria plays host to an annual electronica festival named for Rainbow Serpent.

Crowds at the Rainbow Serpent electronica festival, Australia, 2013.

5 Rainbows in Art and Film

What skilful limner e'er would chuse
To paint the rainbow's varying hues,
Unless to mortal it were given
To dip his brush in dyes of heaven?
Sir Walter Scott[1]

Two-dimensional depictions of rainbows are almost as old as human art itself, with images of Rainbow Snake featuring in Australian Aboriginal paintings that have been dated to between 2000 and 4000 BCE.[2] In European classical antiquity, the painting of rainbows reflected 'prevailing uncertainties about basic colours'.[3] The philosopher Democritus (*c.* 460–*c.* 370 BCE) believed that there were just two primary colours, red and green; and in his lifetime – and far beyond – this idea proved far more popular than his concept that the universe was made up of atoms. Alexander of Aphrodisias, for whom the rainbow was red, green and violet, argued that these three colours could 'neither be procured nor imitated by painters', and 'the notion that the rainbow was unpaintable persisted well into the nineteenth century.'[4] Through most of history, then, attempting to paint one was often a self-conscious act of mastery – or even of hubris – and was usually seen as such by the contemporaries of those artists who dared to try.

Rainbow painting before 1700

This was perhaps less of a problem in the Middle Ages and Renaissance, when most European painting was religious in character (and indeed sponsorship), and any depiction of the rainbow was ipso facto heavily freighted with Judaeo-Christian symbolism. The Archangel Michael is depicted with rainbow wings in the fifteenth-century English cleric John Lydgate's *Siege*

Rainbow mandorla in Giotto's fresco of the *Last Judgement*, Scrovegni Chapel, Padua.

Japanese watercolour of squash, blossom and rainbow, before 1907.

of Thebes, and a bright-pink, peach-pink and lapis rainbow is very prominently placed above Christ in Crispijn van den Broeck's *Het Laatste Oordeel* (*c.* 1570), in an echo of the bright rainbow-striped 'mandorlas' – whole-body halos – of Christian paintings and mosaics from up to twelve hundred years earlier.[5] Notably, these include a mosaic of Christ in the sixth-century Basilica of San Vitale, Ravenna; Giotto's *Last Judgement* in the Scrovegni Chapel in Padua (*c.* 1304); and Hans Memling's *Triptych of St John the Baptist and St John the Evangelist* (*c.* 1479). The other-worldly brightness of the rainbow colours in such works is very much to the fore. Giotto's, for example, contains three shades of orange and two of green, but it is the central, wide, white stripe that stands out and lends the mandorla its luminous appearance. Memling achieved a similar effect using different colours: bright yellow and bright orange fringed with dark green and dark blue. The San Vitale mosaic, curiously, is one of a very few artistic images even loosely to prefigure the Newtonian rainbow colours and their order (being red-brown, red, pink, yellow, green and blue); but again, it is the relative 'glow' of the

107

central pink/yellow that one chiefly remembers about it.[6] Many other medieval rainbow mandorlas can also be seen, and have been called 'paradigmatic' images of heavenly glory.[7]

It is perhaps to be expected that Queen Elizabeth I, who made a concerted and largely successful effort to supplant the Virgin Mary as an object of popular veneration in the sphere of visual culture, should also have been painted with a rainbow on at least one occasion. Known as *Non sine sole iris* or simply 'The Rainbow Portrait', one of the most mysterious and widely debated portraits of the queen shows her grasping the rainbow as if it is a small bow, with which she is about to set off on a hunt, or to war. Where artists in Catholic realms still portrayed the rainbow as a gigantic, other-worldly explosion of colour, surrounding Christ but perhaps beyond even His control, in Protestant hands (literally) it is reduced to the status of a small tool, and specifically, a tool of state violence. Rainbows have also occurred, very occasionally, in European heraldry – notably, in the badge of Queen Catherine de' Medici of France (r. 1547–9) and in the crest of the Edwards-Moss family of Roby Hall, Lancashire.[8]

Painted rainbows that were very bright, especially in their central stripe or stripes, continued to be the norm through the seventeenth century, even in works that were not overtly religious in character: for instance, in Francisco Rizi's equestrian portrait of Marie-Louise d'Orléans of 1679. Three important exceptions should be noted, however. The first was the work of Lucas van Uden (1595–1672), whose naturalistic landscapes were

Maarten van Heemskerck (1498–1574), *Panorama with the Abduction of Helen amidst the Wonders of the Ancient World*, 1535, oil on canvas.

The rainbow-as-bow in a portrait of Elizabeth I, variously attributed to Marcus Gheeraerts the Younger and Isaac Oliver, *c.* 1602.

NON SINE SOLE
IRIS.

often accompanied by strikingly pallid rainbows; he has also been hailed as the first to use the bow in a post-religious way. Second, we must consider van Uden's much more famous associate, Peter Paul Rubens, whose *Landscape with a Rainbow* (1636) – in fairly sharp contrast to the same painter's more traditional and perspectivally flawed *The Rainbow Landscape* of a few years later – depicts a double rainbow of such pale and wispy immateriality that one might reasonably wonder if a fog bow is intended.[9] The third is Jan Siberechts, who in *Landscape with Rainbow, Henley-on-Thames* (*c.* 1690) created another double rainbow of a terrible pallor. Whether this latter picture represents an early artistic rejection of the Newtonian rainbow's visual jollity, or merely an homage to Rubens, is difficult to discern. But certainly, the shapes, relative widths and positions of both Siberecht's bows are well observed, as compared to the prevailing artistic standards of his time.

In the immediate aftermath of Newton's discoveries – perhaps not coincidentally – Nonconformist Protestant theology took a brief turn against the rainbow's association with Christ.

Peter Paul Rubens, *Landscape with a Rainbow*, *c.* 1636, oil on oak panel.

Jan Siberechts,
*Landscape with
a Rainbow,
Henley-on-Thames,*
c. 1690, oil on canvas.

In 1712, perhaps responding to his fellow early New Englanders' preference for having their funerary monuments made 'in every color of the rainbow', Reverend Cotton Mather dismissed as 'fanciful' that the bow – a mere image of the Sun, and an 'imperfect' circle that falls as the Sun rises – was somehow a fitter symbol of the Great Redeemer than the Sun itself. Rather, he held that as a broken, fallen, imitation Sun, the rainbow was a suitable Puritan image for the individual believer who has dedicated himself to imitating Christ's example.[10]

Non-scientists' acceptance of the scientific insight that the rainbow was a false and imperfect Sun was far from limited to Protestants, however. Less than a generation after Mather, a very different commentator – the Roman Catholic poet and translator Alexander Pope – expressed a similarly dismissive attitude, comparing unoriginal fictional characters to 'a mock-rainbow ... the reflexion of a reflexion'.[11] In his private life, Pope was fascinated by prismatic and other optical effects, and boasted in 1725 that the grotto he had built in Twickenham

> becomes on the instant, from a luminous Room, a Camera Obscura, on the walls of which all the objects of the River, Hills, Woods, and Boats, are forming a moving Picture . . . And when you have a mind to light it up, it affords you a very different Scene: it is finished with Shells interspersed with Pieces of Looking-glass in angular Forms . . . at which when a Lamp . . . is hung in the Middle, a thousand pointed Rays glitter and are reflected over the place.[12]

This deep interest in the mingling of sculptural and optical effects arguably reached a peak with Pope's water garden on the same site. Sometimes known as *giochi d'acqua*, such sculptural/optical spectacles became relatively popular in England from the mid-seventeenth century, when Thomas Bushell constructed a set at Enstone in Oxfordshire that would, among many other things, create artificial rainbows and cause 'two fountains of rose-coloured water rise into the air, each suspending a golden ball'.[13] Scientists' oft-repeated reformulations of the rainbow each 'represented an entirely new way of seeing a natural phenomenon, and one that ushered in a major transformation', in the arts as much as in science.[14] This was never more the case than in the transition from the discourse of 'wonders' to that of scientific optics between the mid-seventeenth and mid-eighteenth centuries, as part of which the spectrum became separated once and for all from other spectres of a more traditional and altogether more menacing variety.

Rainbow painting, 1700–1900

But it was Siberechts, not Pope, who had pointed the way forward, at least in the medium term. In contrast to the brightness of the mostly religious rainbows that preceded them, and the numberless army of kitsch ones that would follow, the fine-art rainbows of the eighteenth and nineteenth centuries are characterized by an extraordinary pallor. Only a fraction of this can be ascribed to either deterioration or unavailability of pigments. *The Four Seasons Paying Homage to Chronos* by Bartolomeo Altomonte (d. 1783) prominently features a three-part bow in which the upper and lower bands are both white, and the centre gold, with the dull grey-blue of the sky showing clearly in between each. The same apparent permeability of the bow by the sky behind it is evident in *Noah's Thanksgiving* by Domenico Morelli (1823–1901), which features a primary bow of orange, white and blue and a secondary of green, white and orange. As well as being disconnected from one another, Altomonte's three bands are apparently quite solid, for an aloof blue-clad female figure reclines on them as if on a couch. In Jacob Philipp Hackert's *The Waterfalls at Terni* (1779), a rainbow appears in spray; painted in the exact same shades of pale blue, cream and white as the falls themselves, it would nearly blend in with them, were it not for its own distinctive geometry.

Bartolomeo Altomonte, *The Four Seasons Paying Homage to Chronos*, c. 1737, oil on canvas.

Similarly, Jean Ranc's equestrian portrait of Philip v of Spain of around 1730 contains a rainbow which – at least in its two-banded portion nearest the ground – is barely distinguishable from the blue of the sky and the white of the clouds. Only at the very top of the picture, where the sky is much darker, is the bow revealed as pastel shades of pink, yellow, blue and violet. Nearly the same rainbow was produced by Jacob Cats in 1799, in *Herfst Namiddag Waater:* blue shading into white (against pale grey skies) at the foot, but considerably more distinct bands of red, yellow and blue (against dark purplish clouds) at the crest. A fairly extreme example of this confounding of the rainbow and its background can be seen in one of the most famous rainbow paintings of all, the 'meteorologically impossible' *Salisbury Cathedral from the Meadows* (1831) by John Constable, who long before creating this particular picture 'came to regard the rainbow as a personal emblem'.[15] In Constable's *Hampstead Heath with a Rainbow* (1836), the colours of the purple, white and burnt orange double bow are all mirrored fairly precisely by those of the sky and clouds. And even the strongest rainbows of the early nineteenth century – those painted by Josef Anton Koch in his *Thanksgiving of Noah* (1803) and *Paysages héroiques* (1805 and 1812) – owe their strength to a centre stripe of verdant green, the very darkness of which seems calculated to oppose or reject the brightness of the central stripes of the rainbows in medieval and early modern art. At their upper and lower edges, which are respectively pale pink and sky blue, Koch's bows fade into the sky as much as any others of their own era.

The overall whiteness or greyness of the painted bow in the Age of Reason is often apparent, however, even in cases where contrast between the bow and its background is strong. J.M.W. Turner's magnificent *Buttermere Lake, A Shower* (1798) has as its centrepiece a milky-white rainbow with the merest hint of peach at its crown. Joseph Wright's *Landscape with Rainbow* of 1795 depicts a more clearly bi- or tricolour bow of faint yellow and white or yellowy white above a pale blue-grey; it stands against a sky exactly as dark as Turner's, but in place of Turner's foggy apparition – an energy-neutral rip in the fabric of the sky

John Constable,
*Salisbury Cathedral
from the Meadows*,
1831, oil on canvas.

– Wright's bow glows like a laser beam, and the idea that it is being fired down at the Earth rather than up from it is palpable. Likewise set against a cloudy and threatening sky, the rainbow in Caspar David Friedrich's *Chalk Cliffs on Rügen* (c. 1810) appears at first glance to be entirely white, before slowly revealing itself as three bands, of white, pale yellow and pale blue. In Friedrich's far more dramatic *Mountain Landscape with Rainbow* (1810), the top two bands have merged into the palest of yellows, but one's overall sense of the bow is similar. Like effects are attempted in Károly Markó's *Italian Landscape with Viaduct and Rainbow* (1838), in which the fairly strong and solid-looking white-rimmed yellow of the primary bow contrasts strongly with the almost ghostly grey insubstantiality of the secondary; and in George Inness's *A Shower on the Delaware River* (1891), whose even darker clouds lend a specious whiteness to a rainbow that slowly emerges as – what else? – red, white and blue. Atkinson

Albert Bierstadt,
The Golden Gate, 1900,
oil on canvas.

Grimshaw's *Seal of the Covenant* (1868) features a wispy bow that is mostly yellow, emerging from the cloudy background only by virtue of its hue, not its brightness.[16] And in Albert Bierstadt's *The Golden Gate* (1900), the three-banded bow's two uppermost bands are both nearly white, albeit tinged with the gold of the title, and its lower band a very pale green.

In short, post-Newtonian artists' rejection of the idea that the rainbow consists of seven distinct colours in a particular order appears to have been nearly absolute. Constable's notebooks, in particular, indicate that this decision was deliberate: having weighed a lifetime of observational experience against an equally solid, fundamentally correct understanding of what Newton had proposed, the painter found Newton wanting.[17] Even in Sir John Everett Millais's *The Blind Girl*, an extraordinary double-rainbow painting of 1856, the violet-blue-green portion of both bows appears to be nearly an afterthought, and the overall effect is of tricolour bows completely dominated by their unnaturally wide central stripes of yellow. As he was a leading exponent of the Pre-Raphaelite style of painting, which advocated and practised a return to the Continental styles of the 1400s, the neo-medievalism of Millais's bows is only to be expected.[18]

Joseph Wright,
*Landscape with a
Rainbow*, *c.* 1795,
oil on canvas.

Caspar David
Friedrich, *Mountain
Landscape with
Rainbow*, 1809–10,
oil on canvas.

The reasons that visual artists of the eighteenth and nineteenth centuries would have rejected the Newtonian conception of the rainbow are not far to seek. First, they may have shared in the poets' visceral rejection of the idea that the phenomenon was something completely explicable, as discussed in Chapter Four. But visual artists were also understandably confused due to 'a persuasive body of pseudo-scientific rhetoric' – much of it emanating from Benjamin West – that falsely claimed colour-mixing operated in more or less the same way regardless of whether one was dealing with light or pigments.[19] The highest expression of this confusion may have come with Angelica Kauffmann's painting *Colour* (1778–80), in which the personification of the art of painting reaches up into the sky to dip her paintbrush directly into a rainbow.[20]

In 1810, however, the phoney peace between the two fundamentally different processes of colour-mixing came to an end with the publication of Johann Wolfgang von Goethe's *Theory of Colours*. The summation of a direct attack on Newton that had begun by the early 1790s, Goethe's thesis was arrived at through a study of painting and many thousands of naturally occurring minerals. Despite the polemic against Newton having been omitted from its first English translation, Goethe's book would

George Inness,
*A Shower on
the Delaware River*,
1891, oil on canvas.

L. Prang & Co.
chromolithograph
of Yellowstone Lake
in the 1870s, after
Thomas Moran
(1837–1926).

go on to profoundly influence many painters, including the
Pre-Raphaelites and Turner; other philosophers, including especially Schopenhauer and Wittgenstein; arguably some poets
including Gerard Manley Hopkins; and even the designers of the
flags of the new republics that were supplanting Spanish colonial
rule in Latin America.[21]

To make a long story short, Goethe disputed the primacy of
Newton's physical colours over the physiological and/or psy-
chological colours perceived by the individual, harking back
to a position he had taken in 1772: 'The human being himself,
to the extent that he makes sound use of his senses, is the most
exact physical apparatus that can exist.'[22] On this basis, Goethe
resurrected a long-discredited Aristotelian position that darkness was not a mere absence, but a force equal to and opposing
light. In this view, the colour blue was not light per se, but a
form of darkness, weakened or blunted by the presence of light;
its opposite number – and the only other true colour – was
yellow: light blunted by darkness. Goethe's system, while absurd
from the viewpoint of physics, popularized a six-colour version
of the 'colour wheel' that is still of relevance to the visual arts
today, unlike Newton's over-complicated seven-colour version.
Goethe's work also may have helped to inspire formal investigation, later in the nineteenth century, of what we now know to

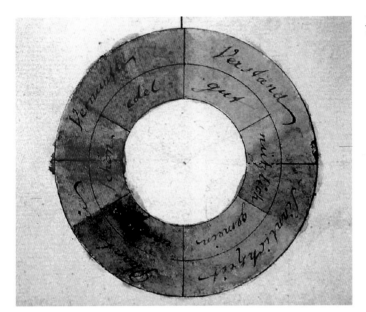

Johann Wolfgang von
Goethe (1749–1832),
Colour Wheel, 1809.

be the fundamental difference between additive mixing and subtractive mixing of colour; that is, that whereas combining the three primary colours of light produces white light (as Newton correctly maintained), combining the primary colours of paint produced black or very dark brown paint (as every painter knew). Newton – anticipating both Pointillist painting and television – was able to create an illusion of pure, uniform whiteness using four different colours of powdered pigment viewed from a particular distance; but if the same pigments had been mixed with a medium such as oil, the same experiment would not have succeeded.

Two-dimensional depictions of rainbows moved inexorably in a kitsch direction over the course of the twentieth century. Though this was very much a 'grassroots' phenomenon, art historians give special mention to Thom Klika of Woodstock, New York. Operating on a fine line between folk art and commercial manufacturing, and often known simply as 'The Rainbow Man', Klika trained as an artist in Ohio before going on to reach millions 'through his books, *Rainbows* and *10,000 Rainbows . . .* his rainbow pocket art, posters, prints, and over 40 years of original rainbow art' including 'rainbow art cards' for the Museum of Modern Art in New York City.[23] There have, however, been a number of noteworthy fine-art uses of the rainbow in the post-Klika environment. One such was John Schroeder's painting *Parable of the Rainbow Dancers* (1975), which features Adam and Eve in the Garden of Eden menaced by space pirates: 'Rainbows in Shroeder's imaginative world . . . turn out to be not just potentially misleading and ambiguous but intentionally so.'[24] But many of the most imaginative recent rainbow artworks have been sculptural and/or dynamic in character, following the lead set by Bushell and Pope centuries ago. At scheduled intervals over twelve weeks in the summer of 2012, Michael Jones McKean's site-specific temporary artwork *The Rainbow: Certain Principles of Light and Shapes between Forms* was presented at The Bemis Center for Contemporary Arts in Omaha, Nebraska.

A dramatic Victorian interpretation of one of Isaac Newton's prismatic experiments.

Rainbow at the Grand
Canyon, before 1913.
Photograph by Arnold
Genthe (1869–1942).

Extensive modifications to the Bemis Center's five-story,
repurposed industrial warehouse took place – creating a
completely self-contained water harvesting and large-scale
storage system. Throughout the project cycle, collected
and recaptured stormwater was filtered and stored in six
above-ground, 10,500 gallon [40,000 l] water tanks. Within
the gallery, a custom designed 60-horsepower [45 kw] pump
supplied pressurized water to nine nozzles mounted to the
20,000 square foot [1,900 m²] roof . . . At timed intervals, in
the morning and early evening, a dense water-wall projected
above the building in which a rainbow emerged.[25]

A new iteration of the work is currently being planned for northern California.

Olafur Eliasson has also created rainbow artworks around the world, including *Your Rainbow Panorama* (2011), a 150 m (490 ft) circular corridor of glass in the colours of the spectrum, mounted above the roof of the AROS Kunstmuseum in Aarhus, Denmark. His 2008 show *Take Your Time* at the PSI Contemporary Art Center in New York included artificial rainbows among a range of 'immersive environments' dedicated to 'probing the cognitive aspects of what it means to see'.[26]

Rainbows in the realm of non-dynamic sculpture are vanishingly rare, but Patrick Hughes's print *Leaning on a Landscape* (1979) plausibly represents one in which a solid rainbow leans against a wall and casts a shadow upon it, thus calling direct attention to the rainbow's status as a non-object. Brilliantly and unusually, postcard pictures of this work add an entire extra level to its principal artistic statement.

'Niagara Rainbows', 1910, lithograph by Joseph Pennell (1857–1926).

Rainbows in film

Motion pictures containing rainbow-like effects are nearly as common as those with night scenes, thanks to lens flare – a generally unwanted phenomenon caused by bright light (such as from the headlamp of a car) being reflected back from the surface of the camera lens, and/or from impurities or imperfections within the material from which the lens is made.[27] The use of zooms tends to worsen this type of interference; lens coatings and lens hoods can help to reduce it, but do not eliminate it entirely in every situation. For digital cameras in particular, a distinct type of rainbow-like artefact can also be caused by diffraction through the image sensor.

Perhaps because celestial rainbows are so fickle, and film production so expensive, real rainbows are seldom written into movie plots. Many films and television programmes with the word 'Rainbow' in the title are merely referring to real-world place names, such as *Tiger Bay and the Rainbow Club* (1960) and *Adventures in Rainbow Country* (1970–71). Many other productions with 'Rainbow' in their names are making commonplace political or quasi-political references to multiculturalism, environmentalism and so forth. 'Don't Stretch the Rainbow', the title of an episode of *21 Jump Street* (1987–91), refers to race relations; and an emotive and controversial feature-length documentary from 1985 about conflicts between and among the Hopi, Navajo, mining interests and the u.s. Government was given the title *Broken Rainbow*. A similar film from 1992, covering the Ho people's (eventually successful) protests against a

Man posing by Yana Zegri's *Evolutionary Rainbow* mural, Haight-Ashbury district, San Francisco, 2012.

Michael Jones McKean, *The Rainbow: Certain Principles of Light and Shapes Between Forms*, Bemis Center for Contemporary Arts, Omaha, Nebraska, 2012.

World Bank-funded hydroelectric project in Bihar State, India, was named *Follow the Rainbow* after one of the protest songs featured.

Undoubtedly the most remarkable transformation of a cinematic work by a single song was the magic worked on Victor Fleming's *The Wizard of Oz* (1939) by 'Somewhere Over the Rainbow'. The classic L. Frank Baum book *The Wonderful Wizard of Oz* (1900), from which the musical film was adapted, makes

it clear that the Land of Oz – like Jonathan Swift's Lilliput, Samuel Butler's Erewhon or Rider Haggard's Kukuanaland – is merely one of the countries of the world, reachable by fairly normal methods of transport. Indeed, far from lying on or beyond the rainbow, the world of Baum's first book makes no mention of rainbows at all, though it is characterized by a rigid colour-coding scheme in which the Munchkins who inhabit the east of the country wear mostly blue; their western counterparts, the Winkies, only yellow; their southern cousins the Quadlings, mostly red; and the people of Kansas, only grey. Fleming's film eliminates the Quadlings, transforms the Winkies into some-thing like Imperial Russian soldiers, and surrounds the Munch-kins in a rainbow array of colours. The film also depicts Glinda, the Good Witch of the North (South, in the book), travelling inside a pulsating ball of ever-changing colours. As well as having been the last word in special effects in its day, Glinda's ball further reinforces the idea that what the audience is seeing is somewhere beyond the rainbow, rather than merely beyond the borders of Kansas. Though the impetus behind the writing

Chris Booth's memorial to the *Rainbow Warrior* above Matauri Bay, New Zealand, where the ship's wreckage was given a traditional Māori burial.

of Baum's thirteen sequels to *The Wonderful Wizard of Oz* was the
runaway success of the Broadway musical version of the original
book in 1902, that show did not include any rainbows either; nor
did the William Selig-produced short film version of 1910. It
was only in the fifth book in the series, *The Road to Oz* (1909),
that the rainbow was briefly introduced, as the father of a 'sky
princess' or 'cloud fairy' by the name of Polychrome, who went
on to appear in the stage play *The Tik-Tok Man of Oz* in 1913, as
well as further Baum novels. But it would be difficult to make a
case that this relatively minor character was influential upon, or
even definitely known to, the songwriter of 'Somewhere Over
the Rainbow'.

Born Isidore Hochberg in Lower Manhattan to Yiddish-
speaking Russian parents in 1896, Edgar 'Yip' Harburg would
grow up to be one of the major lyricists of the twentieth century,
and play 'a major role in the transformation of the Broadway
revue into the sophisticated musical of the 1940s and 1950s'.[28]
Along the way, Harburg wrote the enduring classics 'Brother Can
You Spare a Dime?' (1932) and *Finian's Rainbow* (1947), which
will be dealt with in more detail below. In 'Somewhere Over the
Rainbow', which was eventually named the greatest recording of
the twentieth century by the u.s. National Endowment for the
Arts, the sixteen-year-old Judy Garland, portraying the even
younger child Dorothy Gale, expresses a vague yearning for a
better world than that of the Depression- and Dust-Bowl-era
Midwest. One of the oddest aspects of watching the film hard
upon reading the book is that the former teems with people and
animals in a way that Baum's Kansas of 1900 decidedly does not.
In part, this is due to the film-makers' decision to double-cast
all of the key inhabitants of Oz with Kansas analogues, as part
of an it-was-all-a-dream narrative that the book entirely lacked.
However, this dream strategy is undercut significantly by the
decision to shoot all of the Kansas scenes on black-and-white
film and all of the Oz scenes in Technicolor, which paradoxically
lends the latter – despite their fantastical, over-the-top staginess
– a far stronger air of reality. Coupled with Dorothy's signature
song, the use of colour creates a counter-narrative of actual travel

(beyond the rainbow, in this case) that is far truer to the original book than the film's dialogue, considered in isolation, was ever intended to be.

For reasons that are far from obvious today, 'Somewhere Over the Rainbow' was among the works by Harburg that were considered so blatantly Leftist that he was blacklisted by the House Un-American Activities Committee from 1950 to 1962, and thereby forbidden either to leave the country or to work within it.[29] However, the immediate trigger for Harburg being labelled, or libelled, as a dangerous socialist was *Finian's Rainbow*, a musical play in which a racist senator from the fictional southern state of Missitucky is transformed into a black man, via leprechaun magic imported from Ireland by immigrant Finian McLonergan. Thought to be Broadway's first show with a racially integrated cast, it ran to 725 performances in 1947–8.

The movie version, starring Fred Astaire as McLonergan and Petula Clark as his daughter Sharon, released in 1968, was the first studio feature directed by a twenty-something Francis Ford Coppola. Relentlessly bizarre, and containing small directorial touches that would find echoes in projects as diverse as *Pulp Fiction* and *Police Squad!*, it commences with the footsore McLonergans touring the most iconic beauty spots of the United States, hardly meeting another soul, before arriving in Rainbow Valley County, Missitucky. The economy of the valley revolves around a co-operative enterprise/commune that is applying genetic engineering in pursuit of its own holy grail: tobacco that comes out of the ground ready-mentholated. Apparently pagans, the co-op members at one point perform a maypole dance around a gigantic straw horse, in a clear echo of Hawthorne that is not in any sense motivated by the plot; they are also 'colourblind' in the racial sense. All racism in the community is imposed upon them from above: by the sheriff, senator, district attorney and the white members of those officials' immediate staffs. Even this racist administrative caste is religious only by implication, insofar as they bring a prosecution against Sharon under a witchcraft statute of 1690. And it is primarily this, rather than the racism, that leads Finian (who, early on, had held

Poster for the stage
musical of *The Wizard
of Oz*, *c.* 1902.

America in great esteem) to denounce the country as barbaric
and medieval. Though Sharon ends up marrying a local tobacco
impresario and remaining in Rainbow Valley, the final scene is
of Finian's departure for his native Ireland. His dreams of riches
and assimilation are both in ruins, but he maintains his cheeky
and convivial optimism. He will go on chasing the rainbow, for
'there may not be a pot of gold at the end of it. But there's a
beautiful new world under it.' The implication was clear:

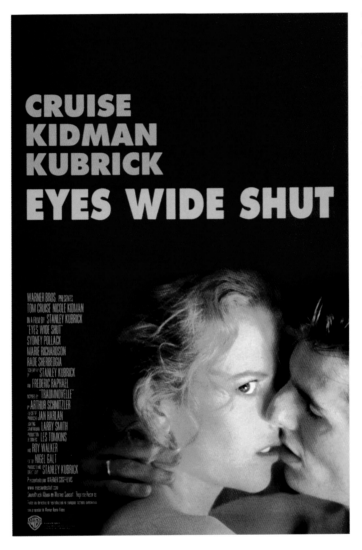

The background colour of the poster advertising the U.S. theatrical release of *Eyes Wide Shut* reinforced the film's association of the main character with the 'bottom' of the rainbow.

America's civil-rights problems were getting better, but the Hollywood of *Star Trek*, *I Spy* and *Guess Who's Coming to Dinner* was standing by to give things a further nudge in the direction of pluralism.

The Wizard of Oz entered the Internet age early and in spectacular fashion, with a widely believed rumour that Pink Floyd's *Dark Side of the Moon* album could be used – and perhaps, had intentionally been written – as an alternative soundtrack for the film of 1939. Like most magical beliefs, this one contained an opt-out clause based on operator error, for no-one was ever quite sure which roar of the MGM lion was the cue to drop the needle for perfect synchronization of the music with the pictures. Regardless of which roar is chosen, however, the results are only mildly and intermittently amusing. Indeed, my experimental screenings with the CD version of the album on 'shuffle' mode beginning from track two have yielded much funnier and more interesting results, including on one occasion Floyd's reference to a paperboy coinciding perfectly with a Munchkin unrolling a scroll marked 'Death Certificate'. On the whole, it seems totally incredible that a group of musicians as experienced as Floyd (who had, in fact, already worked on film soundtracks prior to *Dark Side*'s release) would have produced such a clumsy effort intentionally. But like other Internet rumours and hoaxes, this one has a life all its own, and will probably outlive us all.

Quite unlike natural rainbows, man-made images of rainbows are so commonplace, particularly in urban North America, that they can be hard for film-makers to avoid. Perhaps the most intriguing cinematic use of this latter type of rainbow was in Stanley Kubrick's final, semi-finished masterpiece, *Eyes Wide Shut* (1999). The main character, Bill, is a medical doctor on the lower fringes of Manhattan high society who is driven almost, but not quite, to the brink of self-destruction by jealousy of his wife's flirtations and sexual fantasies. At a party, he is picked up by two attractive young women who clearly intend to seduce him into an adulterous threesome, but they will only go so far as to say they wish to take him to 'where the rainbow ends'. As if disbelieving that they have said it, Bill repeats this phrase and it is repeated back to him several times. But nothing happens; he is called away to a medical emergency. The next night, having baled out of an impromptu assignation with an undergraduate/prostitute, Bill meets an old friend and learns the password to a secret high-end sex party. Desperate to attend, but lacking the necessary costume, he wakes up the owner of a costume shop called Rainbow, which has a brightly lit sign consisting chiefly of a cartoonish Newtonian rainbow. Another shop located directly below it is called Under the Rainbow. During this visit, the shop owner 'catches' a woman he identifies (perhaps falsely) as his underage daughter about to have sex with two cross-dressing Japanese tourists. When Bill returns his hired costume the following day, having been literally unmasked as an interloper at the rich perverts' party, it is clear that the Japanese men have spent the night in the costume shop, which is, on some level or another, also a brothel. Kubrick's naming of the two shops could be dismissed (as a mere joking reference to *The Wizard of Oz*, perhaps) were it not for the persistent association of Bill with the colour violet. He appears through a set of improbable violet-painted wrought-iron gates; is framed by violet light coming in through windows of his flat, among other places; and stands twice outside the student-prostitute's violet front door. Almost everywhere he goes – black-haired, dressed in a black overcoat and black gloves – this colour surrounds him like a halo.[30] No joyous act of adultery, but Bill himself is 'where the rainbow ends'. He, Dr Bill Harford, is the black fringe of the violet bottom of the rainbow's heap of colours, low man on the totem pole; a mere bourgeois who looked upwards at the higher-wavelength colours of his city's super-rich and mistakenly came to believe that being on their spectrum made him an essential part of their society. That he is not actually destroyed by his experiences, or by this realization, is the final measure of his irrelevance.

6 Rainbows in Politics and Popular Culture

Where does the rainbow end, in your soul or on the horizon?
Pablo Neruda[1]

Between 1979 and 1987, under the sponsorship of the East German communist regime, Werner Tübke painted the largest oil painting in the history of the world: more than 120 m (400 ft) long and 14 m (45 ft) high, so large in fact that it required the construction of a special building to house it. Taking as its subject the defeat of Thomas Müntzer's peasant army near Bad Frankenhausen in Franconia on 15 May 1525, it is titled *Frühbürgerliche Revolution in Deutschland* (Early Bourgeois Revolution in Germany). Its content, aside from the immense, lurid and strangely angled rainbow arching directly over the battle, is downbeat and essentially realistic. Unsurprisingly, it disappointed those who had commissioned it, hoping for an unambiguous celebration of the peasants' proto-socialistic heroism.[2] The presence of the rainbow, if not its form, has an obvious basis in reality. According to an eyewitness to the battle named Hans Hut,

> God almighty wanted now to cleanse the world,
> had taken power away from the governing authorities
> and given it to their subjects . . . [The peasants] all
> carried the rainbow as symbols in their banners. This
> rainbow, Müntzer told them, symbolized the covenant
> God had made with them. And after he had preached
> to the peasants continuously for three days, a rainbow
> did in fact appear in the sky. Müntzer drew their attention
> to this, encouraged the peasants and said: You see the
> rainbow now, it is the sign and covenant that God is on

your side. This, Müntzer continued, should encourage them to fight bravely.[3]

But fight bravely they did not; and despite outnumbering the professional forces mustered against them, Müntzer's believers armed with scythes and flails were crushed utterly, and inflicted a mere handful of casualties on the opposing side. In Thomas

Werner Tübke's panorama *Frühbürgerliche Revolution in Deutschland* (Early Bourgeois Revolution in Germany), c. 1976–87.

Nashe's proto-novel *The Unfortunate Traveller* (1594), the hero Jacke Wilton becomes mixed up in the siege of the Westphalian city of Münster in 1535, which had been taken over by radical Anabaptists and transformed into a communistic and polygamous theocracy. Nashe's text suggests that there, too, the victory of the fanatics was mis-foretold by a rainbow, but in the event, they were

135

Statue in Stolberg
of Thomas Müntzer
(1489–1525) waving his
rainbow banner.

empierced, knockt downe, shot thorough . . . [such]
that one could hardly discerne heads from bullettes,
or clottered haire from mangled flesh hung with
gore . . . Heare what it is to be Anabaptists, to bee
puritans, to be villaines.[4]

But whether Nashe's rainbow was part of an otherwise lost oral
tradition, or poetic licence, or merely confusion between Münster
and Müntzer, remains unclear.

The idea that Müntzer's peasants were the first modern revo-
lutionaries, as improbable as it now appears, was part of a long
and respectable intellectual tradition in Germany, beginning
with the monumental historical work of Wilhelm Zimmermann
(1807–1878), published in the mid-1840s. Zimmermann, a disciple
of the German idealist philosopher Georg Hegel (1770–1831),
personally rejected religion.[5] Perhaps inadvertently, he created a
warped portrait of Müntzer and Müntzer's followers that was
drained of religious content, seeing in their anti-authoritarianism
'the greatest and only correct philosophical system devised by
man'.[6] The historian went on to participate in the March Revolution of 1848 and was elected as a deputy to the revolutionary
Frankfurt National Assembly of 1848–9, caucusing with the left
wing there. These views and activities made Zimmermann a
hero to Friedrich Engels (1820–1895), co-founder of Marxism;
and thus, with the benefit of hindsight, Müntzer and his ragged
brethren would become unlikely icons of communism.

Rainbows, however, largely did not; and prior to the Tübke
episode in the 1970s, their appearance in official socialist art was
rare to non-existent.[7] Throughout the first third of the twentieth
century – a period that saw communist revolutions break out
(with varying degrees of success) in Russia, Germany, Hungary,
Finland, El Salvador and Mongolia – the rainbow motif largely
failed to appear, though in China it did feature on the flag of
the pre-communist republic established in 1912. One Western
observer, ignorant of the many mythic resonances between
rainbows and dragons, claimed that

in substituting this new national emblem for the old flag
of the Chinese Empire which displayed a great Dragon
with hungry jaws, the Chinese Republic seems . . . to have
admitted that the days of the swallowing Dragon were
over, and had been succeeded by a division of their land
into strips, symbolizing the swallowing by five foreign
Powers, England, France, Russia, Germany and Japan.[8]

Officially, however, the five 'strips' or stripes on the Chinese
rainbow flag represented – in addition to all five colours of the
rainbow as traditionally conceived there – the country's different
peoples:

the red one standing for those of the original eighteen
provinces of China, the yellow for the Manchus, the blue
(or, more properly, the 'ching') for the Mongolians, the
white for the Thibetans, and the black for the folk of
Chinese Turkestan.[9]

Communist North Korea saw an interesting blending of rainbow
myth and politics, when official biographers of dictator Kim Jong-
il declared that the Supreme Leader's birth on Baekdu Mountain
in 1941 was 'prophesied by a swallow and heralded with a double
rainbow and a new star in the heavens'.[10]

A symbol in flux: the Americas, 1600–1960

Outside China and the British colony of Ceylon (now Sri
Lanka), where it had a religious precedent in the 'rainbow body'
of the Buddha, the main use of the rainbow as a national or quasi-
national symbol was in Latin America.[11] It would be tempting
to suggest that this was a result of Mexico and Peru having been
Christianized beginning in the sixteenth century, when – German
Peasants' War aside – the idea that the biblical rainbow's prom-
ises 'applied to all living creatures and not to a particular people'
meant that the rainbow covenant 'was peculiarly open to Christian adoption, Catholic and Protestant alike'.[12] But according to

Wiphala and Inca rainbow flags in Lima.

Bernabé Cobo, writing in the 1600s, the peoples of the Andes had their own pre-conquest systems of heraldry, in which the *arco celeste* or rainbow was particularly prominent.[13] By the end of the seventeenth century, indeed, the rainbow had become a favoured feature on the coats of arms that were drawn up by the Spanish heraldic authorities for those members of Native American aristocracies who had remained in the conquerors' good books.

This endorsement by Spain did little to diminish the rainbow's power as an Andean symbol of anti-colonialism, however. As of 2006, 48 per cent of Bolivians favoured the adoption of the *wiphala* – a rainbow flag symbolizing indigenous resistance – as 'a formal patriotic symbol' of their country, and a seven-colour version duly became the 'co-official' flag of the country three years later.[14] This, however, was just one of at least half a dozen versions of the *wiphala* to have been created in the preceding hundred years, including one for the Che Guevara-trained Maoist guerrilla force known as Túpac Katari. In all versions of the *wiphala*, the stripes run diagonally, so the horizontally

THE BALLS ARE ROLLING — CLEAR THE TRACK—

striped rainbow flag of the city of Cuzco in Peru is not considered one, despite having a common origin in Inca culture.

In 1938 the u.s. military promulgated a series of 'Rainbow' plans for the defence of the Western Hemisphere, initially only as far south as the 'bulge' of Brazil ('Rainbows' 1–3) and latterly to include the hemisphere in its entirety ('Rainbow' 4). The name of this series of plans was apparently intended to contrast them with single-colour and bicolour strategies of the earlier inter-war period, such as the apocalyptic 'Red-Orange' of 1928 which envisaged the u.s. fighting alone on four fronts against Britain, Canada, Japan and Mexico. 'Rainbow' 4 was never adopted because the Americans lacked sufficient military resources to execute it, not only in the interwar period but even as late as 1944. However, the nomenclature was extended into the war years with the u.s. Navy's 'Pot of Gold', a plan in May 1940 to send 110,000 troops to Brazil to counter a possible pro-Axis coup there.

The reasons for the choice of this imagery for plans relating to Latin America is far from clear. However, the u.s. war planner

An 1856 political cartoon by Nathaniel Currier, showing Millard Fillmore and James Buchanan crushed by giant balls representing the western and northern states' opposition to slavery and Mormonism. Republican Party slogans appear on the rainbow.

Rainbow representing fair trade, January 1907, by J. S. Pughe (1870–1909).

Colonel Charles Furlong had an 'apparently . . . comprehensive knowledge of the history, economy, geography, and culture of Latin America' and therefore would likely have been familiar with the use of the rainbow as a symbol of national, regional and pan-Amerindian solidarity within South America.[15] It might also be explained by the influence of *Rainbow Countries of Central America* (New York, 1926), a well-reviewed book by Wallace Thompson, a Fellow of the Royal Geographical Society, who chose to associate red with the soil of Costa Rica; orange with the morning sky of Nicaragua; and yellow, blue and green with the landscapes of Honduras, Guatemala and El Salvador, respectively, on the basis of direct observation.

The reason for the naming of the u.s. Army's 42nd 'Rainbow' Infantry Division, now associated exclusively with the northeast, was apparently that on its formation in 1917 it included the best regiments from more than two dozen states in every part of the country. Some sources claim that Douglas MacArthur, then the 42nd's divisional chief of staff, suggested the name because of this

Col. Sherrill and Capt. Andrews turning on the Rainbow Fountain at the Lincoln Memorial, October 1924.

Poster for Imperial
Germany's ninth
war loan, 1918, by
Otto Ubbelohde
(1867–1922).

unusually diverse geographical spread. MacArthur, a Protestant of a stern but non-specific sort, was Army Chief of Staff from 1930 to 1937, with wide responsibilities for war planning, and he may have been behind the raft of rainbow associations in war plans of that time. Early versions of the 42nd's divisional badge showed the rainbow as a three-coloured half circle, while later ones were reduced to a quarter-circle, to commemorate the loss of half the unit's personnel to death or wounds in the First World

War. The u.s. Victory Medal for the same conflict, issued retro-
actively from April 1921, was suspended from a five-coloured,
vertically striped double-rainbow ribbon, with violet along the
edges and orange in the centre.

In light of this almost exclusively militarist context by the
interwar years, it is somewhat surprising that the rehabilitation
of the rainbow as a leftist symbol also began in America – see
the discussion of Isidore Hochberg in Chapter Five, and 'Noah's
Legacy? The Rainbow as Peace Sign', below. In any case, as in
Britain and Ireland, the American utopian political movements
had been decried as 'rainbow chasing' since the later eighteenth
century, a dismissal probably based on the scientific truth that

'Honor Flag' awarded
by the u.s. Treasury
Department.
Lithograph by
W. F. Powers Co.,
1917.

Bain News Service photo of Col. William N. Hughes Jr, wearing the second or 'quarter-circle' version of the Rainbow Division's sleeve insignia – revised from a half-circle to represent the unit's battle losses.

rainbows can only be observed, and never actually approached.[16] The symbolism, if any, behind Emanuel Goren's invention of rainbow sherbet in Philadelphia in the 1950s has not been clearly identified.

Rainbows and the European left, 1600–1960

In England, 'Rainbow' occurs as a surname, including of one Colonel Thomas Rainbow, also Rainsborough, a martyred leader of the radical-egalitarian Levellers of the 1640s. As such, it can be exceedingly difficult to sort out whether places in the historical record called the 'Rainbow Tavern' or similar constituted overt political references, or scientific ones, or were mere punning based on the names of owners or frequent visitors. It has been argued that the Tory poet laureate John Dryden (1631–1700) was making a veiled reference to the 2nd Earl of Sunderland – a politically promiscuous and pseudo-pious courtier of the same era – with his reference to 'various Iris', a rainbow-goddess who is particularly 'changeable' in Dryden's translation of Virgil, though not in Virgil's original text.[17] Given the rainbow's history of association with Nonconformist Protestant radicalism, with which Dryden as an ex-Puritan was no doubt familiar, a rainbow image for the Whig Sunderland was especially apt.

u.s. Victory Medal, First World War.

Less than a decade later, London's Rainbow coffee-house, located in Lancaster Court off St Martin's Lane, became an epicentre of feverish intellectual exchange, mostly among French Protestant exiles and English atheists. Though their interests ranged from science, philosophy, theology and religious toleration to journalism, typography, theatre and chess, the net effect of the activity of the Rainbow's customers was political, in that 'all helped to create the climate in which the radical thought of the Enlightenment could develop later in the century'.[18] The philosopher David Hume would briefly live at the Rainbow in 1739.[19] Given that the dominant philosophical ethos of the 'Rainbow Group' was scepticism – the idea that real knowledge is unattainable – the rainbow motif is highly appropriate, even if probably accidental.

A similar phenomenon would occur beginning in 1894, when a leftist splinter

A 'patriotic' rainbow
over New York
appeared in this
wartime poster
targeting food
waste by immigrants.
Lithograph by Charles
Edward Chambers
(1883–1941).

FOOD WILL WIN THE WAR
You came here seeking Freedom
You must now help to preserve it
WHEAT is needed for the allies
Waste nothing

group of the British Liberal Party first met in the Rainbow
Tavern, Fleet Street. Two centuries earlier, this had housed the
shop of Samuel Speed, printer and bookseller, who was arrested
in 1666 for distributing potentially seditious works from the
Cromwell era there.[20] The Victorian group became known as the
'Rainbow Circle', a name that stuck long after they changed
venue to a member's house in Bloomsbury Square. Members of
the all-male circle included future Labour prime minister John

LIBERTY SOWING the SEEDS of VICTORY

Write for Free Books to

NATIONAL WAR GARDEN COMMISSION
WASHINGTON, D.C.

Charles Lathrop Pack, President

P.S.Ridsdale, Secretary

Poster from 1917
by Frank Vincent
DuMond, echoing
President Woodrow
Wilson's dictum 'Food
will win the war.'

Ramsay MacDonald, future Liberal Party leader Herbert Samuel and the visionary anti-imperialist economist J. A. Hobson, who though not a socialist or communist himself became a major influence on the thought of Vladimir Lenin. For nearly four decades, the group – limited by its own rules to thirty members at any given time – served as 'an important intellectual laboratory for . . . the welfare ideology that permeated British social-democratic politics in the early twentieth century'.[21] But it split over the question of whether to support or oppose the First World War, and afterwards never recovered its pre-war stature.

Similar usages could be found in more far-flung parts of the Empire. In 1876 E. W. Cole (1832–1918) established a book arcade which 'at its peak . . . encompassed two blocks from Bourke to Collins Streets in Melbourne, Australia'. The rainbow sign hung at its entrance 'stood for progressive, radical thinking' and for books as 'the means to creating world peace by eradicating ignorance, fear and hatred'.[22]

In post-Peasants' War continental Europe as elsewhere, the rainbow retained its religious symbolism; however, it was largely shorn of any political associations until after the First World War. In the early 1920s, arguably the high water mark of communist feeling in Western Europe, it was adopted as a symbol of the non-communist Left, in the form of the International Co-operative Movement. This was first proposed in Basel in 1921 at a congress of the movement, which held (in contrast to communism) that the goals of socialism could be achieved without revolution. Professor Charles Gide – a prominent French Protestant, 'Co-operator' and Christian Socialist, and uncle of Nobel Prize-winning author André Gide – designed a seven-colour rainbow flag that emblematized the Co-operative concept of 'unity in diversity'; this was formally adopted in 1925.

Noah's legacy? The rainbow as a peace sign, 1800–2000

The British-born American radical pamphleteer Thomas Paine (1737–1809), who had been a privateer and excise officer at various times, proposed while living in Revolutionary France in the

World Peace Flag.

summer of 1800 that flags 'composed of the same colors as compose the Rainbow, and arranged in the same order as they appear in that Phenomenon' be used at sea by ships whose countries were not at war. This was the centrepiece of his 'plan for the protection of the Commerce of Neutral Nations during War'.[23] Though Paine's proposal was not adopted, it was widely read, and can be identified as the first modern, non-religious appearance of the rainbow as a 'peace sign'.

It was next used in this context in 1913, when an American Methodist minister named James van Kirk devised a busy and frankly unattractive 'world peace flag', consisting of a field of white stars on a dark blue ground, on which was superimposed a block of seven horizontal rainbow stripes (red uppermost) on the hoist side, tethered via eight white ropes to a predominately white globe of the Earth on the fly side. This design received considerable attention due to van Kirk's four world tours and extensive merchandising, and the flag was officially adopted by the Universal Peace Congress, an international body that met 33 times between 1889 and the outbreak of the Second World War.

In Italy in 1961 peace activists created a flag consisting of the seven colours of the Newtonian rainbow, violet uppermost, plus an eighth, white stripe above that – presumably as a nod to earlier conventions concerning flags of truce and surrender. This white stripe was later replaced by the superimposed word *pace*, Italian for 'peace', again in white. Bridging the gap between the 1960s international peace movement and later environmentalism, Greenpeace was founded in Vancouver, Canada, in 1969 to protest against U.S. underground nuclear weapons testing in Alaska, which it was feared might cause earthquakes and tsunamis. Arising in part from Quaker anti-war principles and that religion's practice of bearing silent witness to injustice and wrongdoing, the group's original and most famous tactic was ship-to-ship confrontation, usually in international waters. From an exclusively anti-nuclear focus down to 1975, Greenpeace branched out via anti-whaling activities into its current, broad range of ecological pursuits. Its most famous oceangoing craft, the 40 m (130 ft) former trawler *Rainbow Warrior*, was sunk by French commandos in Auckland, New Zealand, in 1985,

Rainbow Warrior II in Turkish waters, 2009.

with the loss of one crewman's life. The replacement vessel, *Rainbow Warrior II*, bore an eight-striped, seven-colour rainbow logo, the two outermost stripes both being white – presumably a reference to peace-movement origins, though a specific legend of the Cree Nation of American Indians is also cited as an origin for the ship's name. One of Greenpeace's most well-publicized and successful interventions took place in 1996, when a group of its members occupied a Shell Oil offshore drilling platform, the *Brent Spar*, to prevent it being disposed of by intentional sinking in the North Atlantic. Among the tactics allegedly used to dislodge Greenpeace members from the platform was constant harassment with fire hoses. In a (thankfully, bloodless) echo of the Battle of Frankenhausen nearly five centuries earlier,

> Suddenly the water cannons just stopped. We walked out onto the platform to see if anything was happening. On the *Altair* [Greenpeace support ship] we could see little silhouetted figures dancing around. We couldn't figure out what was going on, until the *Altair* radioed us and told us the great news [that Shell had agreed not to sink the platform]. After that moment an incredible rainbow appeared in the sky.[24]

Gay Pride, kitsch and the New Left, 1969–2000

The rainbow was not in evidence during the foundational event of the modern Gay Pride movement, the riots centred on the Stonewall Inn on Christopher Street, New York City, in late June and early July 1969. It is widely believed that the six nights of rioting were somehow sparked by the death of Judy Garland on 22 June; historians have failed to find any evidence for this – there were much more immediate triggers including Mafia exploitation and police brutality – though Garland was already an icon for American gay men at the time.[25] In any case, 'Somewhere Over the Rainbow', its socialist political meaning (if any) long since forgotten, became integral to the Gay Pride marches that began on the first anniversary of the Stonewall riots and continue to this day.

'Somewhere over the Rainbow' is interpreted as a wistful fantasy on a life where homosexuals can live openly and fully without judgment, social rancor, and rejection. Ironically, it is both the pathos in Garland's rendition of the song and her self-destructive life (thought to be involuntary) that have made her an icon of contemporary gay life in America.[26]

As such, the song probably played a role in the immediate popularity of the first Gay Pride flag – a rainbow of eight colours, including hot pink, and with turquoise replacing blue – which was created in 1978 by San Francisco artist Gilbert Baker. Within two years, apparently due to the unavailability of sufficient quantities of pink and turquoise fabric, the design had been modified to the present six-colour design, with red uppermost, pink omitted and royal blue replacing both indigo and turquoise. A version of the six-colour Gay Pride rainbow, 1.5 km (1 mile) long and 9 m (30 ft) high, organized by Baker for the 25th anniversary of the Stonewall riots in June 1994, set a record for the longest flag in the history of the world up to that time. Baker himself broke this record nine years later, this time with a 2-km (1¼-mile) flag in the original eight colours. At one point in the late 1970s, demand for rainbow flags was so strong in the San Francisco gay community that it completely outstripped supply;

Rainbow banners decorating the California Capitol Building in celebration of the outcome of Obergefell v. Hodges, 26 June 2015.

in place of 'official' pride items, some residents adopted flags made for the International Order of the Rainbow for Girls, a still-active masonic leadership organization for eleven- to 21-year-olds that had been inaugurated at the Scottish Rite Masonic Temple in McAlester, Oklahoma, in 1922.

Nevertheless, the concept of Gay Pride remained largely ignored in mainstream American society as of 1978, when the American Broadcasting Company launched *Mork and Mindy*: a sitcom spin-off of ratings titan *Happy Days*, with a young Robin Williams reprising his role as space alien Mork from Ork. Mork's trademark costume item, when in disguise as a human, was a pair of rainbow-coloured braces, mass-market imitations of which quickly became popular among children and adults. The following year, *The Muppet Movie* opened with popular puppet Kermit the Frog (performed by Jim Henson) wistfully singing 'The Rainbow Connection', written by Paul Williams and Kenneth Ascher. A near-sequel to 'Somewhere Over the Rainbow', it became unexpectedly popular on the radio, remaining in the U.S. top 40 charts for nearly two months, and was nominated for an Academy Award. At least one version of the film's poster featured a prominent naturalistic rainbow. In 1994 an act called Greg and Steve released an anti-racism song aimed at children, 'The World Is a Rainbow', which became popular across North America; and the hemispherical Waldo Tunnel in northern California, which has long been painted with a rainbow motif, became the subject of a campaign to rename it the Robin Williams Tunnel after that actor's death in 2014.[27]

Kitsch explodes

The early 1980s saw the first comprehensive integration of children's products with children's television programming. The Rainbow Brite universe, devised by Hallmark Cards of Kansas City, Missouri, in 1983, combined television content with books, dolls, dollhouse furniture, other toys and games, branded school supplies, puzzles, jewellery and cosmetics, luggage, clothing, towels, radios, lamps and even bicycles. The content was driven by the struggle of the forces of colour, led by a girl named Rainbow Brite and her rainbow-tailed horse Starlite, against the forces of darkness, led by the King of Shadows and latterly, the Dark Princess and Murkwell Dismal (echoing Nathaniel Hawthorne's 'jollity and gloom . . . contending for an empire').[28] The same blurring of retail merchandise and content – probably traceable to the astonishing success of *Star Wars* action figures from the late 1970s onwards – saw Hallmark's rival American Greetings adapt their existing Care Bears product line-up for series television and feature films, beginning in 1985.[29] Each Care Bear had a 'tummy symbol' or 'belly badge' denoting its personality: a rainbow adorned the stomach of Cheer Bear, one of the original ten characters, all of whom functioned more or less as guardian angels. All Care Bears have heart-shaped noses, but Cheer Bear was unusual in that her heart-nose was red, even though her fur was pink. The My Little Pony franchise was launched in 1981 as a line of Hasbro toys (originally called My Pretty Pony), with a set of Rainbow Ponies added in 1983. The title of the *My Little Pony* television series of 1986 was inscribed on a rainbow in the opening sequence, and the character named Rainbow Dash has remained enduringly popular in a variety of media, including her own straight-to-DVD animated feature. There is also a My Little

Rainbow Brite cosplay.

Pony videogame called *Crystal Princess: Runaway Rainbow*. The latest My Little Pony feature-length film, *Equestria Girls: Rainbow Rocks*, portrays the adventures of a band called the Rainbooms.[30] This refers to the ability of Rainbow Dash, who is the lead guitarist for the band, to fly supersonically and create a sonic boom with a rainbow-coloured shockwave. Through the 1980s and beyond, U.S. viewers of these and other TV programmes would also have been bombarded with rainbow-centred advertisements, notably including the long-running campaigns for Lucky Charms breakfast cereal (featuring a leprechaun) and Skittles candy ('Taste the rainbow'). Video game-maker Taito, more famous for *Space Invaders*, released a game called *Rainbow Islands* in 1987.

Perhaps the oddest rise-and-fall story in the rainbow-kitsch universe has been Lisa Frank, Inc., an Arizona-based retailer focused on school supplies, which reportedly netted its eponymous founder and her husband an average $10 million a year between 1995 and 2005. Criticized as 'the world's shittiest employer' and a 'rainbow gulag', the company allegedly forbade its workers from speaking to one another and secretly taped their phone calls; the penalties for violations of its increasingly bizarre rules 'ranged from verbal abuse to name-calling to screaming

Newsagent's shop,
Arndale Centre,
Manchester.

to automatic termination'. In the words of one of a legion of disgruntled former employees, this was 'kind of ironic, given that they have rainbows and unicorns everywhere'.[31]

Inevitably, given the sheer quantity of rainbow kitsch that had been generated by the 1990s, parodies of it became more numerous. Memorable examples included the video for the Republic of Ireland's fictive Eurovision Song Contest entry on the hit Channel 4 sitcom *Father Ted*;[32] *Saturday Night Live* comedian Jimmy Fallon singing the twee theme song from PBS educational series *Reading Rainbow* in the signature style of Doors lead singer Jim Morrison; the parodic video game *Robot Unicorn Attack*, whose titular antagonist sports a rainbow mane and tail; a high-school musical called *God and His Magical Rainbow Suspenders* that was created as part of Fox's hit animated series *Family Guy*; and the 'Pleasure Town' animated sequence used in place of a sex scene in the 2004 comedy film *Anchorman*. The ongoing popularity of rainbow kitsch items is so strong, however, that it is often impossible to tell whether a particular one is intended parodically or not. For instance, a clumsily executed animation of a smiling pink and grey cat that leaves a six-striped rainbow trail as it flies across the sky has been viewed more than 130 million times on YouTube.

Most intriguingly, the notorious underground-comics artist R. Crumb produced a 'straight' illustrated version of the entire biblical Book of Genesis in 2009; it was nominated for a total of five Harvey and Eisner awards, winning one. An extreme statement on the power of illustration alone to shift the meaning of a text in unexpected directions, its rainbow passages – to which Crumb arguably devotes disproportionate attention – have been criticized for focusing on Noah's 'stupidity or confusion' and God's particular displeasure at the sin of murder, as distinct from displeasure at mankind's sinfulness in a more general sense.[33]

Rainbows in coalition

Popular connotations of generic niceness merged neatly with the early twentieth-century socialist theme of 'unity in diversity' in

Rainbow PUSH
Coalition members
protest the
impeachment of
President Bill Clinton,
December 1998.

1984, when African American Baptist preacher Jesse Jackson described his U.S. presidential campaign as the product of a multi-ethnic 'rainbow coalition' that included the disabled, small farmers, working mothers, the unemployed, gays and lesbians, the young and anyone else who felt they had been harmed by the policies of the Ronald Reagan administration.[34] Soon formalized as the National Rainbow Coalition, the organization still exists, with headquarters in Chicago and branches in eight other cities. Though the extent to which Reverend Jackson's appeal in fact reached beyond the black community has been questioned, the campaign itself was hailed as

> a watershed event in the history of black politics . . . perhaps on a par with the August 1963 march on Washington, the passage of the Voting Rights Act of 1965, and the election of the first black mayors in 1967.[35]

More than any other single event, this sealed the rainbow's status as a symbol of both cross-ethnic cooperation and of the

Left in North America. By 1992 the 'rainbow coalition' phrase had caught on in the Republic of Ireland, where it was associated with

> social urbanisation, cultural modernisation and institutional Europeanisation . . . encapsulated in a new type of nationalism (if that it even be) more affirmative and self-confident, sceptical of clerical authority, bridling at under-performance and corruption, and . . . a belief that the 'civil war' parties no longer measure up to the challenges facing the state – particularly, mass unemployment.[36]

Two years later, the term was applied semi-officially to the three-party left-wing ruling coalition led by John Bruton, which remained in power in Ireland until 1997.

Postcolonial Africa

To the disgust of many, the white-minority regime in South Africa would use the old Co-operative slogan 'Unity in Diversity' in 1981, for the celebration of the twentieth anniversary of the country's separation from the British monarchy and Commonwealth.[37] The retention of this saying after South Africa's re-entry into the Commonwealth with the coming of democracy in 1994 may have played a role in the widespread rhetoric of the 'rainbow people' and 'rainbow nation', made famous by Archbishop Desmond Tutu and President Nelson Mandela, respectively, in a neat blending of the themes of multiculturalism and peace. In 1994 Tutu famously began describing South Africans as 'rainbow people of God', a reference both to the Bible story of Noah as well as to the multiple tribes and ethnic groups making up modern South Africa. Mandela borrowed this notion and altered it somewhat, describing the post-Apartheid country as 'a rainbow nation, at peace with itself and with the world'.[38] In that same year, the u.s. Agency for International Development convened a meeting in Harare, Zimbabwe with representatives of all Anglophone African countries, plus Britain, Jamaica and

Nepal, for the purpose of assessing the 'potential of radio in Africa as a tool for . . . development'. Among other things, this quickly resulted in a Nigerian radio drama called *Rainbow City*, focused on 'issues of democracy and good governance' and fully sponsored by the u.s. Information Service during its first three months on the air. Written in 'a combination of local and national pidgins and conventional English all garnished with local or indigenous ingredients such as proverbs, idioms, [and] puns', *Rainbow City* was hailed as a great success, being broadcast twice a week throughout the Islamic north of Nigeria and on sixteen stations elsewhere in the country, and supported by a 'dependable' network of listeners' clubs; and its popularity seems to have played a small but important role in Nigeria's relatively peaceful transition from military to civilian rule in 1999. According to one of its producers, the name of the show recognizes 'diversity':

> Most of the action takes place in a typical large block of rundown flats ['Endurance Villa'] . . . home to a mix of characters that deal with the everyday problems of survival and of getting along with each other.[39]

Anti-rainbow backlash

Use of the universal Buddhist rainbow flag that had been introduced in Ceylon in the mid-1880s was banned in South Vietnam in 1963 by the brutal Catholic-minority government of President Ngô Đình Diệm. Shootings of flag-ban protestors led directly to widely publicized self-immolation protests by Buddhist clergy that in turn destabilized the country, leading to the assassination of Diệm and a catastrophic decade-long u.s. intervention in the region.

In post-Apartheid South Africa, critiques of the Tutu/Mandela 'rainbow people' and 'rainbow nation' rhetoric began almost immediately, with many on the Left – particularly non-blacks – dismissing it as 'candy-coated myth' and a 'flimsy illusion'.[40] But this was at the end of the twentieth century, the high point of the diffusion, and therefore vagueness, of the

rainbow as a political symbol, simultaneously indicating the co-operative, peace, gay rights, environmentalist and multiculturalist movements, alone or in any combination.[41] This slightly absurd situation probably reached its nadir in the Westminster parliamentary election campaign of George Weiss, also known as Rainbow George, of the Vote for Yourself Party in 2001. Formerly known as Captain Rainbow's Universal Party, it was part of the British 'rainbow alliance' of non-serious politicians that also included the Monster Raving Loony Party and the Beer, Fags and Skittles Party.[42] Contesting all four seats in Belfast, which he hoped would be renamed 'Belfeast' as part of the reunification of the UK with the Republic of Ireland under the name 'Emerald Rainbow Islands', Weiss advocated the replacement of parliament with electronic referenda and renaming the pound sterling the 'wonder'. Each Emerald Rainbow citizen would receive a fixed salary of 27,000 wonders a year, and be forbidden from accumulating more than 1 million wonders. The penny would be called a 'gasp'.[43]

Julita Wójcik's rainbow sculpture in Saviour Square, Warsaw, following an arson attack in 2013.

Since that colourful and unedifying time, however, the rainbow has been increasingly regarded as a sort of gay juggernaut. In 2000, the athletics programme of the University of Hawaii at Manoa dropped the rainbow from its logo and team names, allegedly to distance them from homosexuality. The following year, the International Cooperative Association suspended use of their original rainbow flag from 1925, citing its similarity to other well-known flags, and replaced it with a white flag charged with a quarter-circle rainbow splintering into a flock of rainbow-coloured doves. But even this dramatically shrunken rainbow proved too gay for some, and it has already been superseded by a plain purple flag bearing the word 'coop' in white.

In Warsaw's Saviour Square in June 2012, there appeared a large rainbow arch consisting of a metal frame decorated with some 16,000 plastic flowers. This sculpture had been previously set up in Brussels during Poland's presidency of the European Union, 'without an overt political message beyond the need for inclusiveness'.[44] Some Warsaw residents hailed its arrival as a 'new symbol of the Polish capital's increasing tolerance'; others tried to burn it down – at least four times in the first year – for, regardless of the original meaning ascribed to it by artist Julita Wójcik and by the staid government cultural department that sponsored her work, a very large number of Poles choose to see the rainbow arch as nothing other than pro-LGBT propaganda, with one telling an Anglo-Polish defender of the new monument that it was a 'provocation' comparable to 'install[ing] a big penis in front of the Westminster Cathedral'.[45] Such comments by the religious Right have led to a backlash by other religious conservatives, many of whom insist that the rainbow be seen – exclusively – as representative of God's covenant with Noah.[46] British embassies worldwide were banned from flying the rainbow flag during Gay Pride events by then-foreign secretary Philip Hammond in 2015, implying that they had commonly done so in the past; and *The Telegraph* called it 'a fitting coincidence' that natural rainbows appeared over Dublin and Cork while the results of Ireland's historic gay-marriage referendum were being tallied.[47]

Most parties to this debate seem to have failed to notice that by its existence, their argument proves that the rainbow indeed means very different things to different people. It probably always will. But satirists Chris Harper and Paul Bradley struck a nerve with their suggestion that the (real) town of Rainbow City, Alabama, has given up the struggle and is contemplating renaming itself. 'The Rainbow was a symbol of beauty, hope, promise, growth and prosperity,' the town's mayor supposedly told the pair in 2013. 'But now it's just gay.'[48]

Perhaps, perhaps. But as a compelling spectacle in culture as much as in nature, and a unique bridge between subjective experience and objective reality, the rainbow's story is far from over.

REFERENCES

1 What Rainbows Are and How They Work

1 Television presents an interesting parallel to this, insofar as only one pixel of a television screen is lit up at any one instant. The 'picture' exists only in the mind of the viewer, who in order to perceive a coherent television image must literally 'connect the dots' on a screen that is, at all times, more than 99 per cent dark.

2 *The Hitchhiker's Guide to the Galaxy* (London, 1979), p. 3. It is also possible that '42' commemorated Adams's big break in television: acting the part of a surgeon in episode 42 of the BBC's acclaimed *Monty Python's Flying Circus*; by its episode 45, he had been taken on as a writer. Adams explicitly rejected the rainbow and all similar explanations, however, insisting that this 'completely ordinary' number simply came into his head as he was typing – though this hardly proves that the rainbow was not involved in his thought process. See also Douglas Adams, *The Original Hitchhiker Radio Scripts*, ed. Geoffrey Perkins (London, 1985).

3 D. Roberson, I. Davies and J. Davidoff, 'Color Categories Are Not Universal: Replications and New Evidence from a Stone-age Culture', *Journal of Experimental Psychology: General*, LXXIX/3 (2000), pp. 369–98.

4 Ari Ben-Menahem, *Historical Encyclopedia of Natural and Mathematical Sciences, Volume 1: Pre-science–1583 CE* (Berlin, 2009), p. 1931; see also Raymond L. Lee, Jr and Alistair B. Fraser, *The Rainbow Bridge: Rainbows in Art, Myth, and Science* (University Park, PA, 2001), p. 236.

5 Specifically, this occurs because 'large raindrops flatten as they fall but smaller ones do not': Lee and Fraser, *Rainbow Bridge*, p. 324.

6 Ben-Menahem, *Natural and Mathematical Sciences*, p. 1917.

7 Except in the case of light rays that are exactly perpendicular to the water's surface, which change speed but not direction.

8 Adapted from 'Snell's Law', www.physicsclassroom.com, accessed 15 January 2016.

9 V. Ostdiek and D. Bord, *Inquiry into Physics*, 7th edn (Boston, MA, 2013), p. 377.

10 Rachel Kaufman, 'Pictures: First Quadruple Rainbow Ever Caught on Camera', 8 October 2011, http://news.nationalgeographic.com, accessed 27 April 2017.

11 Harald Edens, 'Observations of the Quinary Rainbow', www.weatherscapes.com, accessed 14 January 2016.

12 Les Cowley, 'Supernumerary Rainbows', *Atmospheric Optics*, www.atoptics.co.uk, accessed 22 January 2016.

13 A good explanation of this aspect of Young's work, and its drawbacks, is provided in Philip Laven, 'Ray Tracing with Interference (Young's Method)', 11 May 2003, www.philiplaven.com, accessed 27 April 2017.

14 That is, '0.7mm mean dia. with only an 8% (std. dev.) spread in diameters': Cowley, 'Supernumerary Rainbows'. Thomas Young suggested an even smaller size, of 250 to 500 μm, in his *A Course of Lectures on Natural Philosophy and the Mechanical Arts* (London, 1807), vol. I, p. 470.

15 Les Cowley, 'Red Rainbows', *Atmospheric Optics*, www.atoptics.co.uk, accessed 22 January 2016.

16 Les Cowley, 'Plate Orientation and Plate Halos', *Atmospheric Optics*, www.atoptics.co.uk, accessed 24 January 2016.

17 Lisa Winter, 'Fire Rainbows and How They Form', IFL *Science*, www.iflscience.com, accessed 3 January 2016.

18 Les Cowley, 'Supralateral and Infralateral Arcs', *Atmospheric Optics*, www.atoptics.co.uk, accessed 24 January 2016.

19 BBC Wiltshire, '"Rare" Upside Down Rainbow Spotted in Wiltshire', BBC, www.bbc.com, 19 May 2014.

20 BBC *Magazine*, 'Who, What, Why: How Common Are Upside-down Rainbows?', BBC, www.bbc.co.uk, 18 January 2016.

2 Rainbows in the History of Scientific Enquiry

1 Quoted in M.L.G. Redhead, 'Review: Wave-particle Duality', *British Journal for the Philosophy of Science*, XXVIII/1 (1977), p. 65.

2 See Carl B. Boyer, 'Robert Grosseteste on the Rainbow', *Osiris*, XI (1954), p. 247.

3 Raymond L. Lee, Jr and Alistair B. Fraser, *The Rainbow Bridge: Rainbows in Art, Myth, and Science* (University Park, PA, 2001), p. 106.

4 Aydin M. Sayili, 'The Aristotelian Explanation of the Rainbow', *Isis*, xxx/1 (1939), p. 65.

5 Gloria Saltz Merker, 'The Rainbow Mosaic at Pergamon and Aristotelian Color Theory', *American Journal of Archaeology*, lxxi/1 (1967), p. 81.

6 Aristotle, *De caelo*, quoted in Sayili, 'Aristotelian Explanation', p. 68.

7 Sayili, 'Aristotelian Explanation', p. 71. This fact was temporarily obscured by an early nineteenth-century mistranslation of the word for 'reflection' that Aristotle used: ibid.

8 Aristotle, *De anima*, quoted in Sayili, 'Aristotelian Explanation', p. 72. Aristotle also considered, but explicitly rejected, the idea that light travels or has a speed: ibid., pp. 72–3. See also Lee and Fraser, *Rainbow Bridge*, p. 105.

9 We cannot be sure of this; though both men were alive in the last quarter of the fourth century bce, Euclid's life-dates are largely speculative.

10 In fairness, the tertiary and quaternary bows were not caught on camera until 2011: Jason Palmer, '"Quadruple Rainbow" Caught on Film for the First Time', bbc, 6 October 2011, www.bbc.co.uk, accessed 27 April 2017; Rachel Kaufman, 'Pictures: First Quadruple Rainbow Ever Caught on Camera', http://news.nationalgeographic.com, accessed 27 April 2017.

11 This argument of course subverts, or is subverted by, the idea that the observer is at the centre of a hemispherical dome and that the rainbow is projected on the dome's interior wall – which, if true, would place all stripes of the rainbow equidistant from the observer's eye, despite their varying altitudes vis-à-vis the surface of the Earth.

12 These ideas are set forth in Aristotle's *Meteorologica*: see Sayili, 'Aristotelian Explanation', pp. 73–4.

13 Aristotle, *Meteorologica*, quoted ibid., p. 74.

14 Lee and Fraser, *Rainbow Bridge*, p. 102.

15 Homer, for instance, called it a *sideron ouranon* (inverted metal bowl); and Hesiod, several centuries later, referred to the 'dome of heaven': N. L. Geisler and W. D. Watkins, *Worlds Apart: A Handbook on World Views*, 2nd edn (Eugene, or, 2003), pp. 219–20.

16 Specifically, Posidonius held that the cloud acted as a single continuous reflector rather than a medium for thousands of tiny flat mirrors: I. G. Kidd, *Posidonius*, vol. ii: *The Commentary, Testimonia and Fragments 1–149* (Cambridge, 1988), pp. 124–5. See also Boyer, 'Grosseteste on the Rainbow', p. 248.

17 G. D. Williams, *The Cosmic Viewpoint: A Study of Seneca's 'Natural Questions'* (Oxford, 2012), p. 70.

18 Williams, *Cosmic Viewpoint*, p. 71, italics in original.

19 Quoted in Williams, *Cosmic Viewpoint*, p. 75, n. 71.

20 Lee and Fraser, *Rainbow Bridge*, pp. 109, 138–9. Saint Isidore of Seville, for instance, accepted a quasi-Aristotelian explanation for the rainbow, but ascribed it to Pope Clement i: ibid., p. 139.

21 Boyer, 'Grosseteste on the Rainbow', p. 248; Lee and Fraser, *Rainbow Bridge*, pp. 36, 141.

22 R. H. Pierce, 'The Rainbow Mosaic at Pergamon and Aristotelian Color Theory', *American Journal of Archaeology*, LXXII/1 (1968), p. 75.

23 Quoted in Lee and Fraser, *Rainbow Bridge*, p. 139.

24 Ibid.

25 Ibid., p. 142. It should be noted that Chinese explanations for the rainbow remained fundamentally mythological at this date, relying on the concepts of yin and yang: Sayili, 'Aristotelian Explanation', p. 83.

26 Lee and Fraser, *Rainbow Bridge*, p. 146.

27 Boyer, 'Grosseteste on the Rainbow', p. 250.

28 D. C. Lindberg, 'Roger Bacon's Theory of the Rainbow: Progress or Regress?', *Isis*, LVII/2 (1966), p. 240, n. 15 and 16.

29 Ibid., p. 238.

30 Philip Fisher, *Wonder, the Rainbow, and the Aesthetics of Rare Experiences* (Cambridge, MA, and London, 1998), p. 16. I would, of course, categorically dispute Fisher's statement that the nearness he describes 'was always, whether in mythology or science, understood': ibid.

31 Boyer, 'Grosseteste on the Rainbow', p. 251.

32 'His treatise on the rainbow undoubtedly was widely read during the thirteenth and fourteenth centuries, for half a dozen manuscript copies of the work are extant in libraries at Madrid, Oxford, Florence, Groningen, Prague, and the Vatican': Boyer, 'Grosseteste on the Rainbow', p. 251.

33 Ibid., pp. 250–51, 254–6.

34 Lee and Fraser, *Rainbow Bridge*, p. 160.

35 Boyer, 'Grosseteste on the Rainbow', p. 252, n. 15.

36 Ibid., p. 254; for the colours themselves, see Lee and Fraser, *Rainbow Bridge*, p. 159.

37 Lindberg, 'Bacon's Theory', p. 236.

38 Boyer, 'Grosseteste on the Rainbow', p. 254; Lindberg, 'Bacon's Theory', p. 237.

39 J. H. Bridges, ed., *The Opus majus of Roger Bacon* (London, 1900), vol. II, p. 187.

40 Lindberg, 'Bacon's Theory', p. 242.

41 Ibid., p. 236. 'A similar idea is found in Avicenna, but never had it been developed to the level found in Bacon's discussion': ibid., p. 244.

42 Ibid., p. 246.

43 A. C. Crombie, *The History of Science from Augustine to Galileo* [1959] (New York, 1995), p. 124; Theodoric quoted in Lee and Fraser, *Rainbow Bridge*, p. 163.

44 Lee and Fraser, *Rainbow Bridge*, p. 164, brackets and italics in original.

45 Ibid., p. 162; similarities between the work of Theodoric and Qutb have generally been counted as a 'curious coincidence' or the result of a shared intellectual descent from al-Haytham, rather than as a result of direct influence in either direction: Crombie, *Augustine to Galileo*, p. 124.

46 Lee and Fraser, *Rainbow Bridge*, p. 166. For a far less positive view of Theodoric's contributions, see R. C. Dales, *The Scientific Achievement of the Middle Ages* [1973] (Philadelphia, PA, 1994), pp. 94–5.

47 In *Commentari sopra la storia e le teorie dell'ottica* (Bologna, 1814), vol. I, pp. 149–80. For the view that Theodoric's work had been 'overlooked for hundreds of years' prior to 1814, see Dales, *Scientific Achievement*, p. 94.

48 Carl Boyer, 'Kepler's Explanation of the Rainbow', *American Journal of Physics*, XVIII/6 (1950), p. 360; R. W. Wood, *Physical Optics* (New York, 1928), p. 342, quoted in Boyer, 'Kepler's Explanation', p. 360.

49 For instance, Leonard Digges (b. 1520) and William Gilbert (1544–1603): see Lee and Fraser, *Rainbow Bridge*, pp. 170 and 174–6; Boyer, 'Kepler's Explanation', p. 360.

50 Quoted in Lee and Fraser, *Rainbow Bridge*, p. 171.

51 Crombie, *Augustine to Galileo*, p. 124; Boyer, 'Kepler's Explanation', p. 360.

52 R. E. Ockenden, 'Marco Antonio de Dominis and His Explanation of the Rainbow', *Isis*, XXVI/1 (1936), p. 45.

53 Pierre Duhem, 'History of Physics', in *Essays in the History and Philosophy of Science*, ed. and trans. Roger Ariew and Peter Barker (Indianapolis, IN, 1996), p. 172.

54 Boyer, 'Kepler's Explanation', p. 366, which also advances the speculation that the Theodorician explanation could have been passed even beyond de Dominis's lifetime, via a (now lost) treatise written by Willebrord Snellius (1580–1626) in the last decade of his life.

55 All quotations in this paragraph are from Boyer, 'Kepler's Explanation', p. 361.

56 Kepler's *Paralipomena* (Frankfurt, 1604), cited in Boyer, 'Kepler's Explanation', p. 362.

57 Boyer, 'Kepler's Explanation', p. 363; Lee and Fraser, *Rainbow Bridge*, p. 355, n. 189.

58 He also believed that the radius of the secondary bow was 56
    degrees: Boyer, 'Kepler's Explanation', p. 365.
59 C. B. Boyer, 'Descartes and the Radius of the Rainbow', *Isis*, XLIII/2
    (1952), p. 95.
60 Ockenden, 'Marco Antonio de Dominis', p. 45. For the
    commonplace view that Descartes *must* have seen de Dominis's
    work, see for instance Crombie, *Augustine to Galileo*, p. 124.
61 Boyer, 'Descartes and the Radius of the Rainbow', p. 95.
62 Lee and Fraser, *Rainbow Bridge*, p. 184.
63 Karoly Simonyi, *A Cultural History of Physics*, trans. David Kramer
    (Boca Raton, FL, 2012), p. 229.
64 Robert Hooke, *An Attempt to Prove the Motion of the Earth from
    Observations* (London, 1674), p. 2.
65 See Lee and Fraser, *Rainbow Bridge*, p. 186.
66 V. Ostdiek and D. Bord, *Inquiry into Physics* (7th edn, Boston, MA,
    2013), p. 377.
67 Simonyi, *Cultural History of Physics*, p. 228.
68 Lee and Fraser, *Rainbow Bridge*, p. 193.
69 Ibid., p. 192.
70 J. L. Epstein and M. L. Greenberg, 'Decomposing Newton's
    Rainbow', *Journal of the History of Ideas*, XLV/1 (1984), p. 116.
71 Lee and Fraser, *Rainbow Bridge*, p. 195.
72 See especially ibid., pp. 198–9.
73 A. E. Shapiro, ed., *The Optical Papers of Isaac Newton* (Cambridge,
    1984), vol. I, p. 51.
74 Lee and Fraser, *Rainbow Bridge*, pp. 204, 206.
75 Epstein and Greenberg, 'Decomposing Newton's Rainbow', p. 117.
76 'Thus Newton failed in his quest to find a general mathematical
    relationship between color and refraction': Lee and Fraser, *Rainbow
    Bridge*, p. 202.
77 R. H. Stuewer, 'Was Newton's "Wave-particle Duality" Consistent
    with Newton's Observations?', *Isis*, LX/3 (1969), p. 393.
78 Robert Hooke, *Micrographia, Or Some Physiological Descriptions
    of Minute Bodies Made with Magnifying Glasses . . .* (London,
    1665). Wave theory was further elaborated by Christiaan Huygens
    (1629–1695) in his *Traité de la lumière* (Leiden, 1690) and would
    eventually supplant the particle theory when it was found that the
    latter could not explain diffraction. However, the 'corpuscularianism'
    of Newton and Descartes was also adhered to by Pierre Gassendi,
    Thomas Hobbes, John Locke, Robert Boyle and others. None of
    this stopped Robert Watt from arguing in favour of Aristotelian
    whole-cloud rainbow production as late as 1819.
79 A. Lande, 'The Decline and Fall of Quantum Dualism', *Philosophy
    of Science*, XXXVIII/2 (1971), p. 221. The idea that Newton himself

arrived at an early version of this 'wave-particle duality' theory has also been robustly attacked: see Stuewer, 'Newton's "Wave-particle Duality"', pp. 392, 394, and R. H. Stuewer, 'A Critical Analysis of Newton's Work on Diffraction', *Isis*, LXI/2 (1970), pp. 188–205.

80 Lee and Fraser, *Rainbow Bridge*, p. 220.

3 Rainbows and Myth

1 See especially Bernhard Lang, 'Non-Semitic Deluge Stories and the Book of Genesis: A Bibliographical and Critical Survey', *Anthropos*, LXXX (1985), pp. 605–16. On the Gran Chaco tribes mentioned, see Alfred Métraux, *Myths of the Toba and Pilagá Indians of the Gran Chaco* (Phildaelphia, PA, 1946), p. 29.

2 A. R. Radcliffe-Brown, 'The Rainbow-serpent Myth of Australia', *Journal of the Royal Anthropological Institute*, LVI (1926), p. 25; see also Franz Boas, 'The Limitations of the Comparative Method of Anthropology', *Science*, IV/103 (1896), pp. 901–8.

3 Genesis 9:8–17.

4 W. W. Hallo and W. Kelly Simpson, *The Ancient Near East: A History* (New York, 1971), pp. 35–6.

5 Matthew 5:5.

6 J. G. Melton and M. Laumann, eds, *Religions of the World: A Comprehensive Encyclopedia of Beliefs and Practices*, 2nd edn (Santa Barbara, CA, 2010), p. 1412.

7 Madabusi Subramaniam, *At the Feet of the Master* (New Delhi, 2002), p. 104.

8 Sten Konow, *Bashgali Dictionary: An Analysis of Col. J. Davidson's Notes on the Bashgali Language* (New Delhi, 2001), p. 33.

9 For an overview of the concept of 'proto-Indo-European' culture, see J. P. Mallory and D. Q. Adams, *Oxford Introduction to Proto-Indo-European and the Proto-Indo-European World* (London, 2006).

10 R. S. Buswell, Jr and D. S. Lopez, Jr, eds, *The Princeton Dictionary of Buddhism* (Princeton, NJ, 2013), p. 586.

11 Or at any rate arrows, in the case of Peru, the bow being merely implied. See D. Dauksta, 'From Post to Pillar: The Development and Persistence of an Arboreal Metaphor', in *New Perspectives on People and Forests*, ed. E. Ritter and D. Dauksta (Dordrecht, 2011), p. 108.

12 University of Chicago, 'Digital Dictionaries of South Asia: Samsad Bengali–English Dictionary', http://dsal.uchicago.edu/dictionaries/, pp. 138, 508 and 912, accessed 17 July 2014.

13 S. M. Zwerner, *The Influence of Animism on Islam: An Account of Popular Superstitions* (New York, 1920), p. 158.

14 The rarely used alternatives *qaws al-matar* (rain-bow), *qaws Allah* (bow of God) and *qaws al-alwan* (bow of colours) also imply a bow

of the same type; however, the root 'QZH' can refer to something high, or to the mixing of colours in order to make something more beautiful: Gilbert Ramsay, University of St Andrews, personal communication.

15 Raymond L. Lee, Jr and Alistair B. Fraser, *The Rainbow Bridge: Rainbows in Art, Myth, and Science* (University Park, PA, 2001), pp. 9–14.

16 Ibid., pp. 7, 21.

17 Robert Blust, 'The Origin of Dragons', *Anthropos*, XCV (2000), p. 529.

18 Barbara Ambros, 'Vengeful Spirits or Loving Spiritual Companions? Changing Views of Animal Spirits in Contemporary Japan', *Asian Ethnology*, LXIX/1 (2010), pp. 55–7.

19 R. T. Christiansen, 'Myth, Metaphor, and Simile', *Journal of American Folklore*, LXVIII/270 (1955), pp. 417–27.

20 Jay Crain and Vicki Pearson-Rounds, 'A Fallen Bat, a Rainbow, and the Missing Head', *Indonesia*, 79 (2005), p. 68.

21 Lee and Fraser, *Rainbow Bridge*, pp. 31–2. This is not intended as an exhaustive list.

22 'Fairer-than-a-Fairy', in Andrew Lang, ed., *Yellow Fairy Book* (London, 1956), pp. 143–51.

23 Rev. C. M. Chen, 'On Padmasambhava's Rainbow Body', ed. Stanley Lam, 31 October 2000, www.yogichen.org, accessed 17 July 2014.

24 D. C. Ahir, *Buddhism in Modern India* (New Delhi, 1991), p. 180.

25 Blust, 'Origin of Dragons', pp. 527, 532.

26 John Mason, 'Yorùbá Beadwork in the Americas: Òrìṣà and Bead Color', *African Arts*, XXXI/1 (1998), pp. 28–35 and 94.

27 Ibid., p. 29.

28 A. F. Roberts, 'Chance Encounters, Ironic Collage', *African Arts*, XXV/2 (1992), pp. 54–63 and 97–8.

29 Ibid., p. 58.

30 Marta Moreno Vega, 'The Candomble and Eshu-Eleggua in Brazilian and Cuban Yoruba-based Ritual', in P. C. Harrison et al., eds, *Black Theatre: Ritual Performance in the African Diaspora* (Philadelphia, PA, 2002), pp. 153–66.

31 N. L. Norman and K. G. Kelly, 'Landscape Politics: The Serpent Ditch and the Rainbow in West Africa', *American Anthropologist*, new ser., CVI/1 (2004), pp. 98–110.

32 Ibid., pp. 103, 107.

33 John Loewenstein, 'Rainbow and Serpent', *Anthropos*, LVI (1961), pp. 31–40.

34 Blust, 'Origin of Dragons', p. 524.

35 P. Schebesta, *Die Bambuti-Pygmäen vom Ituri* (Brussels, 1950), p. 156, quoted in Loewenstein, 'Rainbow and Serpent', p. 35.

36 R. Brumbaugh, 'The Rainbow Serpent on the Upper Sepik', *Anthropos*, LXXXII (1987), pp. 26, 28, 33. See also M. Mead, 'The Mountain Arapesh II: Supernaturalism', *Anthropological Papers of the American Museum of Natural History*, XXXVII/3 (1940), p. 392; Radcliffe-Brown, 'Rainbow-serpent Myth', p. 24.
37 Radcliffe-Brown, 'Rainbow-serpent Myth', p. 22.
38 R. H. Barnes, 'The Rainbow in the Representations of Inhabitants of the Flores Area of Indonesia', *Anthropos*, LXVIII (1973), p. 612.
39 Loewenstein, 'Rainbow and Serpent', pp. 31–2.
40 Ibid., p. 33.
41 Radcliffe-Brown, 'Rainbow-serpent Myth', p. 22.
42 David McKnight, *People, Countries, and the Rainbow Serpent: Systems of Classification among the Lardil of Mornington Island* (Oxford, 1999), p. 243.
43 Blust cites current beliefs in the Lanzhou area of Gansu Province, China, and among the Muria people of eastern India, as well as 'the Chinese classics': 'Origin of Dragons', pp. 525, 527; Lee and Fraser, *Rainbow Bridge*, p. 30.
44 Loewenstein, 'Rainbow and Serpent', pp. 32, 33; Radcliffe-Brown, 'Rainbow-serpent Myth', pp. 19–20.
45 Radcliffe-Brown, 'Rainbow-serpent Myth', p. 20.
46 The Masai are one of the few groups said to have killed their rainbow serpent: Loewenstein, 'Rainbow and Serpent', p. 35; Blust, 'Origin of Dragons', p. 532.
47 R. H. Heelas, 'Review of *Under the Rainbow: Nature and Supernature among the Panare Indians* by Jean-Paul Dumont', *American Anthropologist*, new ser., LXXXII/4 (1980), p. 897.
48 Blust, 'Origin of Dragons', p. 526; E. H. Spicer, *The Yaquis: A Cultural History* (Tucson, AZ, 1980), p. 64, quoted in Blust, 'Origin of Dragons', p. 527.
49 A view with which Blust concurred: 'Origin of Dragons', pp. 525–6; Loewenstein, 'Rainbow and Serpent', p. 36.
50 See for instance Radcliffe-Brown, 'Rainbow-serpent Myth', p. 24.
51 Lee and Fraser, *Rainbow Bridge*, pp. 22–31.
52 Claude Lévi-Strauss, *From Honey to Ashes: Introduction to a Science of Mythology*, trans. John Weightman (New York, 1973), vol. II, p. 80; but see also Douglas Taylor, 'The Dog, the Opossum and the Rainbow', *International Journal of American Linguistics*, XXVII/2 (1961), pp. 171–2. Neither Blust nor Lee and Fraser take any account of rainbow-skunk or rainbow-opossum lore.
53 Loewenstein, 'Rainbow and Serpent', pp. 37–8.
54 Ibid., pp. 38–40, but see also Grafton Smith, *The Evolution of the Dragon* (Manchester, 1919) and Norman and Kelly, 'Landscape Politics', p. 105.

55 Based on a remark by Radcliffe-Brown in 'Rainbow-serpent Myth', p. 25; Blust, 'Origin of Dragons', p. 519.

56 Ibid.

57 Ibid., pp. 522–3, 533–4.

58 Daniel MacCannell, 'Cultures of Proclamation: The Decline and Fall of the Anglophone News Process, 1460–1642', PhD dissertation, University of Aberdeen (2009).

59 Lee and Fraser, *Rainbow Bridge*, p. 25; 'Bow, *n* 1', Oxford English Dictionary, www.oed.com, accessed 13 July 2016. An arch or vault was a *stán-boga* – literally 'stone-bow' – while *boga*, by itself, was apparently not used at all: ibid.

60 Blust, 'Origin of Dragons', p. 530.

61 A study proposed, but not executed, in Christiansen, 'Myth, Metaphor, and Simile', p. 422, and in Michael Haldane, 'The Translation of the Unseen Self: Fortunatus, Mercury and the Wishing-hat', *Folklore*, CXVII/2 (2006), p. 179.

62 M. H. Mullin, '*People of the Rainbow: A Nomadic Utopia* by Michael I. Niman [review]', *American Ethnologist*, XXVI/2 (1999), p. 509.

63 Ibid., p. 510.

64 Ibid.

65 Christiansen, 'Myth, Metaphor, and Simile', p. 417.

66 F. J. Cheshire, 'Rainbow Magic', *Folklore*, XII/4 (1901), pp. 479–80; Lee and Fraser, *Rainbow Bridge*, especially pp. 27–9.

67 T. Bane, *Encyclopedia of Fairies in World Folklore and Mythology* (Jefferson, NC, 2013), p. 213.

68 Lee and Fraser, *Rainbow Bridge*, p. 28.

69 Ibid., p. 27.

70 Isidore of Seville (d. 636), the Venerable Bede (d. 735) and Rabanus Maurus (d. 856), cited in Lee and Fraser, *Rainbow Bridge*, pp. 36–7.

4 **Rainbows in Literature, Poetry and Music**

1 Thomas Campbell, 'To the Rainbow' (1819), in Malcolm D. McLean, 'Poems to the Rainbow by Campbell and Heredia', *Hispanic Review*, XVIII/3 (1950), pp. 261, 262.

2 '*Epic of Gilgamesh*, Tablet XI, Column IV', MythHome, www.mythome.org, accessed 22 September 2014.

3 M. Haldane, 'The Translation of the Unseen Self: Fortunatus, Mercury and the Wishing-hat', *Folklore*, CXVII/2 (2006), p. 179.

4 E. B. Gilman, '"All Eyes": Prospero's Inverted Masque', *Renaissance Quarterly*, XXXIII/2 (1980), p. 226.

5 N. Frye, *A Natural Perspective* (New York and London, 1965), pp. 157–8.

6 I am thinking especially of William Drummond's Sonnet v, ll. 9–14: 'How sun posts heaven about, how night's pale queen / With borrowed beams looks on this hanging round, / What cause fair Iris hath, and monsters seen / In air's large fields of light, and seas profound, / Did hold my wand'ring thoughts, when thy sweet eye / Bade me leave all, and only think on thee.' Fleetwood Mac's 1987 single 'Seven Wonders' treads essentially the same ground.

7 John Milton, *Comus: A Mask Presented at Ludlow Castle*, in Milton, *Poems upon Several Occasions*, ed. Thomas Warton (London, 1791), poem at pp. 137–263, quotation at pp. 177–8.

8 Henry Vaughan, *The Rain-bow*, in Vaughan, *The Works in Verse and Prose Complete*, vol. 1, ed. Alexander B. Grosart (n.p., 1871), poem at pp. 234–6, quotation at p. 234.

9 George P. Landow, 'Rainbows: Problematic Images of Problematic Nature', The Victorian Web, www.victorianweb.org, accessed 21 January 2016.

10 Quotation at Percy Bysshe Shelley, 'To a Lady with a Guitar', l. 73, in *The Poetical Works of Percy Bysshe Shelley, Edited by Mrs Shelley* (London, 1874), p. 304. A date of April 1822 is tentatively assigned to the poem by the Bodleian Library, which knows it under the alternative title 'With a Guitar, to Jane [Williams]': 'Shelley's Ghost: Reshaping the Image of a Literary Family', Bodleian Libraries, http://shelleysghost.bodleian.ox.ac.uk, accessed 24 October 2014.

11 Percy Bysshe Shelley, 'The Cloud', part v, ll. 5–14, ibid., p. 259.

12 Percy Bysshe Shelley, *Queen Mab*, part VII, ll. 229–31 and 234, ibid., poem at pp. 1–19, quotation at p. 15,. part 7, ll. 229–31 and 234.

13 Jerome McGann, 'Byron, George Gordon Noel, sixth Baron Byron (1788–1824)', Oxford Dictionary of National Biography, www.oxforddnb.com, accessed 23 October 2014.

14 *The Bride of Abydos*, canto II, part XX, ll. 34–9, in *The Poetical Works of Lord Byron Complete in One Volume* (New York, 1867), p. 96.

15 *The Complete Poetical Works of William Wordsworth*, ed. Henry Reed (Philadelphia, PA, 1837), p. 27.

16 Quoted in Geoffrey Durant, *William Wordsworth* (Cambridge, 1969), p. 108. Cf. also Shelley's 'When the cloud is scattered / The rainbow's glory is shed', ll. 3–4 in 'When the Lamp Is Shattered'.

17 'Hymn before Sunrise, in the Vale of Chamouny', ll. 82–3, in *Select Works of the British Poets, in a Chronological Series from Falconer to Sir Walter Scott* (Philadelphia, PA, 1848), poem at pp. 536–7, quotation at p. 537.

18 *The Two Founts*, or *Addressed to a Lady on Her Recovery from a Severe Attack of Pain*, ll. 17–24, in *Quarterly Review*, XXXVII (1828), poem at pp. 90–91, quotation at p. 91.

19 A notion critiqued by Robert Browning (1812–1889) in his poem
'Caliban upon Setebos' (1864); this quotation is from Browning's
editor F.E.L. Priestley. See Marc R. Plamondon, ed. Representative
Poetry Online, http://rpo.library.utoronto.ca, accessed 25 October
2014; see Coleridge, 'Granville Penn and the Deluge – Rainbow'
(1824) in H. N. Coleridge, ed., *Specimens of the Table Talk of the Late
Samuel Taylor Coleridge* (London, 1835), vol. I, p. 51.

20 *Lamia*, part II, ll. 229–38, in *Select Works of the British Poets, in a
Chronological Series from Southey to Croly* (Philadelphia, PA, 1845),
poem at pp. 562–7, quotation at p. 567.

21 Quoted in Eric L. Haralson, ed., *Encyclopedia of American
Poetry: The Nineteenth Century* (New York and London, 1998),
p. 342. Claims in this direction have been made on behalf of
Poe, Nathaniel Hawthorne, Washington Irving and even James
Fenimore Cooper, but including these authors might tend to
stretch the definition of 'Romanticism' to breaking point.

22 Mark Twain, *A Tramp Abroad* [1879] (New York, 1921), p. 170.

23 Henry David Thoreau, *Walden and Other Writings* [1854] (New York,
1937), p. 195.

24 For a useful overview see Walter Blair, 'Color, Light, and Shadow
in Hawthorne's Fiction', *New England Quarterly*, XV/1 (1942),
pp. 74–94, especially p. 91.

25 Nathaniel Hawthorne, *Twice-told Tales* (Philadelphia, PA, 1889),
pp. 49–62. All quotations from this story are taken from this edition.

26 Nathaniel Hawthorne, *The Scarlet Letter: A Romance* [1850]
(Leipzig, 1852), p. 103.

27 Nathaniel Hawthorne, *The House of the Seven Gables* [1851]
(Rockville, MD, 2008), p. 79.

28 M. Nimetz, 'Shadows on the Rainbow: Machado's "Iris de la
Noche"', *Hispania*, LIX/1 (1976), p. 54. The full text of the poem is
reproduced on p. 50.

29 Robert Browning, *The Poems of Browning*, vol. III: *1847–1861*, ed.
John Woolford, Daniel Karlin and Joseph Phelan (Harlow, 2007),
p. 65.

30 Grenville Kleiser, *Dictionary of Proverbs* (New Delhi, 2005), p. 155.

31 E. M. Forster, *Howard's End*, ed. Douglas Mao [1910] (New York,
2010), p. 150.

32 D. H. Lawrence, *The Rainbow*, ed. Mark Kinkead-Weekes
(Cambridge, 1989), p. 181.

33 Ibid., p. 459.

34 Review by Arthur Ramírez, *Hispania*, LXXIII/3 (1990), p. 671; see
Chapter Six.

35 George Barlow, 'Awushioo: Henry Dumas at the Rainbow Sign',
Black American Literature Forum, XXII/2 (1988), p. 168.

36 Shange was born Paulette Williams in New Jersey, the daughter
of a surgeon in the u.s. Air Force; perhaps coincidentally, London's
first mainstream stage play with an all-black cast was Errol John's
Moon on a Rainbow Shawl (1958): Kate Dorney, 'Banbury, (Frederick
Harold) Frith (1912–2008)', Oxford Dictionary of National
Biography, www.oxforddnb.com, accessed 10 September 2014.

37 Quoted in Jill Cox-Cordoba, 'Shange's "For Colored Girls" has
Lasting Power', cnn online, 21 July 2009, http://edition.cnn.com,
accessed 27 April 2017.

38 Paulette S. Johnson, 'When Every Cloud is Drained of Sorrow and
Anger', *Callaloo*, 11 (1978), p. 139.

39 Jonathan Enright, 'The Rainbow', *Books Ireland*, 229 (2000),
pp. 75–6.

40 George MacDonald, *The Golden Key and Other Stories* (Grand
Rapids, mi, and Cambridge, 2000), p. 2.

41 Ibid., p. 35.

42 Clifford Mills, *Where the Rainbow Ends, a Fairy Story by Clifford
Mills based on the Fairy Play of the same name by Clifford Mills and
John Ramsey* (London, n.d. [after 1910]), pp. 18–19.

43 Ibid., p. 15.

44 Ibid., pp. 246–7.

45 'Where the Rainbow Ends', Wikipedia, https://en.wikipedia.org,
accessed 26 January 2016.

46 'Jacob ter Veldhuis Program Notes', Peermusic Classical,
www.peermusicclassical.com, accessed 26 January 2016.

47 Quoted in Inge van Rij, *The Other Worlds of Hector Berlioz: Travels
with the Orchestra* (Cambridge, 2015), p. 146.

48 Mark DeVoto, 'The Debussy Sound: Colour, Texture, Gesture',
in *The Cambridge Companion to Debussy*, ed. Simon Trezise
(Cambridge, 2003), pp. 179–96.

49 Scott Floman, 'Ritchie Blackmore's Rainbow (Polydor '75)',
Scott's Rock and Soul Album Reviews, http://sfloman.com,
accessed 26 January 2016.

5 Rainbows in Art and Film

1 Sir Walter Scott, 'Marmion', canto vi, part v, ll. 20–23, in *The
Poetical Works of Walter Scott, Esq.* (Edinburgh, 1820), vol. iv, p. 155.

2 Raymond L. Lee, Jr and Alistair B. Fraser, *The Rainbow Bridge:
Rainbows in Art, Myth, and Science* (University Park, pa, 2001),
pp. 22–3.

3 John Gage, *Color and Culture: Practice and Meaning from Antiquity
to Abstraction* (Berkeley and Los Angeles, ca, 1993), p. 31.

4 Ibid., pp. 31, 108.

5 British Library MS Royal 18 D ii, fo. 161v.

6 Titian, in the sixteenth century, did occasionally produce 'bows of great complexity, sometimes with upwards of six colours', but this was quite exceptional: Gage, *Color and Culture*, p. 95.

7 Slobodan Curcic, 'Divine Light: Constructing the Immaterial in Byzantine Art and Architecture', in B. G. Wescoat and R. G. Ousterhout, eds, *Architecture of the Sacred: Space, Ritual and Experience from Classical Greece to Byzantium* (Cambridge, 2012), p. 313.

8 Gage, *Color and Culture*, p. 95; J. Fairbairn, *Fairbairn's Crests of the Families of Great Britain and Ireland*, ed. Laurence Butters (Poole, 1986), plate 126d.

9 Though the flaws in question affect nearly everything in the picture of *c.* 1636, the rainbow itself has been criticized as seeming 'to swoop out of the distance on the left and then arc over the foreground trees on the right'; in other words, Rubens 'treats the bow as a solid object that is oblique to the picture plane': Lee and Fraser, *Rainbow Bridge*, p. 124. A discussion of Rubens's various failings in this regard is provided in Gage, *Color and Culture*, pp. 95–6.

10 J. D. LaFountain, 'Colorizing New England's Burying Grounds', in A. Feeser et al., eds, *The Materiality of Color: The Production, Circulation, and Application of Dyes and Pigments, 1400–1800* (Farnham, 2012), pp. 13, 17, 19, 20.

11 'Preface', in Alexander Pope, ed., *The Works of Shakespear* (London, 1728), p. 2.

12 Pope to Edward Blount, 1725, quoted in 'Alexander Pope's Grotto: A Source of Inspiration and Contentment, 1720–1742', The Twickenham Museum, www.twickenham-museum.org.uk, accessed 17 January 2016.

13 Chris Koenig, 'Garden Display Was a Marvel of the Age', *Oxford Times* (27 October 2010), www.oxfordtimes.co.uk, 28 April 2017; see also J. D. Hunt, *Garden and Grove: The Italian Renaissance Garden in the English Imagination, 1600–1750* (Philadelphia, PA, 1986), p. 138.

14 M. D. Garber, 'Chymical Wonders of Light: J. Marcus Marci's Seventeenth-century Bohemian Optics', *Early Science and Medicine*, X/4 (2005), p. 478.

15 P. D. Schweizer, 'John Constable, Rainbow Science, and English Color Theory', *Art Bulletin*, LXIV/3 (1982), pp. 425–7. At the time he painted his first rainbow picture, in 1812, Constable seems to have been unaware that the secondary bow was thicker than the primary or that its colours were reversed.

16 'One is . . . somewhat jarred to discover that Grimshaw has chosen to call his work *The Seal of the Covenant*, thereby claiming

a religious significance for the scene before us which it does not seem to warrant': George P. Landow, 'Rainbows: Problematic Images of Problematic Nature', The Victorian Web, www.victorianweb.org, accessed 21 January 2016.

17 Schweizer, 'John Constable, Rainbow Science, and English Color Theory', p. 431.

18 With William Holman Hunt and Dante Gabriel Rossetti.

19 Schweizer, 'John Constable, Rainbow Science, and English Color Theory', pp. 433, 435–7.

20 That she is extracting colour from a real rainbow, rather than painting a rainbow into the sky (as some have claimed), seems clear from the fact that the only colour on her palette is brown.

21 I am thinking especially of Hopkins's description of the seven-hued rainbow as 'Ending in sweet uncertainty 'twixt real hue and phantasy', in *Il Mystico* (1862), ll. 121–2; World Heritage Encyclopedia, 'Flag of Venezuela', www.gutenberg.us (n.d., 2014 or later), accessed 22 January 2016.

22 J. W. von Goethe, 'Scientific Studies', in D. Miller, ed., *Goethe: The Collected Works* (Princeton, NJ, 1995), vol. XII, p. 57.

23 Rainbow Man, 'Fourth of July Memorial Rainbow Man Celebration, Oxford, Georgia', CNN iReport, 3 July 2009, http://ireport.cnn.com, accessed 28 April 2017.

24 Landow, 'Rainbows: Problematic Images'.

25 'Overview: Michael Jones McKean, The Rainbow', *The Rainbow*, www.therainbow.org, accessed 2 December 2015.

26 'Olafur Eliasson, Take Your Time', MOMA, www.moma.org, accessed 27 January 2016.

27 Intentional lens flare was a major visual characteristic of J. J. Abrams's feature film version of *Star Trek* (2009) for example.

28 Ernie Harburg and Harold Meyerson, *Who Put the Rainbow in The Wizard of Oz?* (Ann Arbor, MI, 1993).

29 Denise Oliver Velez, 'Blacklisting the Rainbow', *Daily Kos*, 1 June 2014, www.dailykos.com, acessed 28 April 2017.

30 Based on Kubrick's customary methods, none of this could be accidental. The director 'would obsess over every visual element that would appear in a given frame, from props and furniture to the color of walls and other objects': 'Cinematographer Larry Smith Helps Stanley Kubrick Craft a Unique Look for *Eyes Wide Shut*, a Dreamlike Coda to the Director's Brilliant Career', The American Society of Cinematographers, www.theasc.com, accessed 28 January 2016.

6 Rainbows in Politics and Popular Culture

1 *The Book of Questions*, trans. William O'Daly (Port Townsend, WA, 2001), p. viii.

2 'Monumentalbild', Panorama Museum, www.panorama-museum. de, accessed 25 August 2014.

3 Paraphrased in Abraham Friesen, 'Philipp Melanchthon (1497– 1560), Wilhelm Zimmermann (1807–1878) and the Dilemma of Müntzer Historiography', *Church History*, XLIII/2 (1974), p. 176. The flags were probably designed by Philipp Goetzgerodt of Mühlhausen: Douglas Miller, *Armies of the German Peasants' War, 1524–26* (Oxford, 2003), p. 45.

4 See https://ebooks.adelaide.edu.au. I am grateful to Prof. Philip Schwyzer for this reference.

5 Friesen, 'Dilemma of Müntzer Historiography', p. 180.

6 Ibid., p. 178.

7 It appears in at least one work of the 1970s by V. G. Sevastianov, but not in any overtly political context.

8 C. H. Sherrill, 'The Five Stripes of China's Flag', *North American Review*, CCXI/773 (1920), p. 517. 'Among the Chinese and Japanese these five hues are considered to comprise all the colors of the rainbow, for in the one which the Chinese call "ching" is included blue, green, purple, and all their shades.'

9 Ibid., p. 517.

10 D. Demetriou, 'Kim Jong-Il: Double Rainbows, Fear of Flying and Godzilla – 10 Things You Might not Know', *The Telegraph* (19 December 2011), www.telegraph.co.uk, accessed 28 April 2017.

11 See Chapter Three. The Ceylonese rainbow flag of 1886, with five broad vertical stripes and five wide horizontal ones in the same colours, was adopted as the flag of world Buddhism in 1950. The second, third and fourth stripes are always yellow, red and white, with the other two varying by country and sect.

12 R. Graziani, 'The "Rainbow Portrait" of Queen Elizabeth I and Its Religious Symbolism', *Journal of the Warburg and Courtauld Institutes*, XXXV (1972), pp. 247–59.

13 B. Cobo, *History of the Inca Empire*, ed. and trans. Roland Hamilton (Austin, TX, 1979), p. 246.

14 J. Dunkerly, 'Evo Morales, the "Two Bolivias" and the Third Bolivian Revolution', *Journal of Latin American Studies*, XXXIX/1 (2007), p. 136: under Article 6 of the Bolivian Constitution of 2008.

15 Col. J. Child, 'From "Color" to "Rainbow": U.S. Strategic Planning for Latin America, 1919–1945', *Journal of Interamerican Studies and World Affairs*, XXI/2 (1979), pp. 237, 246–50, 254.

16 The phrase seems to have been coined by the anonymous author
  of *The Origin of Duty and Right in Man, Considered* (London, 1796),
  but see also H. H. Quint, 'Gaylord Wilshire and Socialism's First
  Congressional Campaign', *Pacific Historical Review*, XXVI/4 (1957),
  pp. 327, 336, and M. Kenneally, 'Autobiographical Revelation in
  O'Casey's "I Knock at the Door"', *Canadian Journal of Irish Studies*,
  VII/2 (1981), p. 35. The phrase is also used in political contexts in
  modern India.

17 Alan Roper, 'Dryden, Sunderland, and the Metamorphoses of a
  Trimmer', *Huntington Library Quarterly*, LIV/1 (1991), pp. 61, 64.

18 E. Grist, 'Rainbow Coffee House Group (act. 1702–1730)', Oxford
  Dictionary of National Biography, www.oxforddnb.com, accessed
  21 November 2013. The Rainbow coffee-house itself existed from
  1702 to 1755.

19 J. Robertson, 'Hume, David (1711–1776)', Oxford Dictionary of
  National Biography, www.oxforddnb.com, accessed 23 October
  2014.

20 S. Wright, 'Speed, Samuel (bap. 1633, d. 1679?)', Oxford Dictionary
  of National Biography, www.oxforddnb.com, accessed 10 September
  2014. Samuel Speed was the grandson of John Speed, the notable
  cartographer and historian.

21 M. Freeden, 'Rainbow Circle (act. 1894–1931)', Oxford Dictionary
  of National Biography, www.oxforddnb.com, accessed 21 November
  2013.

22 J. Grant, '"The Dreams that You Dare to Dream": Rainbows and
  Utopia', http://paintingrainbows.net (12 December 2013), accessed
  5 September 2014.

23 Thomas Paine, 'Maritime Compact, Article 8', found within
  Thomas Paine, 'XXXIII. [Seventh Letter] To the People of the
  United States', in *The Writings of Thomas Paine, Vol. 3: 1791–1804*,
  ed. M. D. Conway (New York and London, 1895).

24 Quoted in B. Kershaw, 'Ecoactivist Performance: The Environment
  as Partner in Protest?', *Drama Review*, XLVI/1 (2002), p. 122.

25 D. Carter, *Stonewall: The Riots that Sparked the Gay Revolution*
  (New York, 2004), especially p. 260.

26 G. L. Davis, '"Somewhere over the Rainbow . . .": Judy Garland
  in Neverland', *Journal of American Folklore*, CIX/432 (1996), p. 127.

27 Jason MacCannell, Special Assistant for Research to Gov. Edmund
  G. Brown, Jr. of California, personal communication via Facebook
  message, 11 September 2014.

28 Nathaniel Hawthorne, 'The May-pole of Merry Mount', in
  Twice-told Tales (Philadelphia, PA, 1889), p. 49.

29 Via occasional TV specials, beginning in 1983. The product line was
  launched in 1981.

30 Unlike several other My Little Pony films, this was released
    in North American cinemas, on 27 September 2014.
31 Tracie Egan Morrissey, 'Inside the Rainbow Gulag: The
    Technicolor Rise and Fall of Lisa Frank', *Jezebel*, http://jezebel.com
    (12 December 2013), accessed 28 April 2017. I am very grateful to
    Jessica Wheelis for this reference.
32 The final five seconds of the one-minute video consisted of the
    titular horse's disembodied head surrounded by a pulsating rainbow
    mandorla; it is available at www.youtube.com.
33 See for instance L. Borders, 'R. Crumb's *The Book of Genesis
    Illustrated*: Biblical Narrative and the Impact of Illustration',
    B.A. thesis, Liberty University (2014), pp. 20–21. I am grateful to
    Prof. Dean MacCannell for having made me aware of Crumb's
    original publication and the above-mentioned oddity of its rainbow
    content.
34 Jesse Jackson, 'The Rainbow Coalition', speech to the Democratic
    National Convention, San Francisco, California, 18 July 1984.
35 G. Loewenstein and L. T. Sanders, 'Bloc Voting, Rainbow
    Coalitions, and the Jackson Presidential Candidacy: A View from
    Southeast Texas', *Journal of Black Studies*, XVIII/I (1987), p. 95.
36 R. Wilson, 'Rainbow Undimmed', *Fortnight*, 312 (1992), p. 5.
37 The saying's adoption as the European Union's motto in 2000
    was decidedly less controversial.
38 *Cape Times*, 'Editorial: Nelson Mandela', IOL, www.iol.co.za
    (10 December 2013).
39 A. A. Martins, 'Radio Drama for Development: ARDA and the
    "Rainbow City" Experience', *Journal of African Cultural Studies*,
    XVI/I (2003), pp. 96, 97 and 99.
40 N. Valji, 'Creating the Nation: The Rise of Violent Xenophobia
    in the New South Africa', unpublished master's thesis, York
    University (2003), p. 26; J. Van Der Riet, 'Triumph of the Rainbow
    Warriors: Gender, Nationalism and the Rugby World Cup', *Agenda:
    Empowering Women for Gender Equity*, 27 (1995), p. 101. See also
    A. Habib, 'South Africa: The Rainbow Nation and Prospects for
    Consolidating Democracy', *African Journal of Political Science/Revue
    Africaine de Science Politique*, II/2 (1997), pp. 15–37.
41 The original, fifteenth-century sense of the adjective 'diffuse'
    was '[c]onfused, distracted, perplexed; indistinct, vague, obscure,
    doubtful, uncertain': 'Diffuse, *adj*', Oxford English Dictionary,
    www.oed.com, accessed 13 July 2016.
42 P. Chippindale, 'Sutch, David Edward (1940–1999)', Oxford
    Dictionary of National Biography, www.oxforddnb.com, accessed
    23 October 2014.
43 M. Byrne, 'Over the Rainbow', *Fortnight*, 396 (2001), p. 8.

44 H. Kozlowska, 'Rainbow Becomes a Prism to View Gay Rights',
   New York Times (21 March 2013), www.nytimes.com.
45 Michael Dembinski, 'Two Rainbows', W-WA Jeziorki
   (8 May 2014), http://jeziorki.blogspot.co.uk.
46 Hrafnkell Haraldsson, 'Christofascist Organization Lays Claim
   to "Gay" Rainbow', *PoliticusUSA*'s Archives, http://archives.
   politicususa.com (16 December 2010), accessed 27 May 2017.
47 Michael Wilkinson, 'Don't Fly Gay Pride Flag, Philip Hammond
   Tells British Embassies', *The Telegraph* (16 June 2015), www.
   telegraph.co.uk; Rosa Prince and Colin Freeman, 'Irish Gay
   Marriage Referendum Ends in Overwhelming Victory for Yes
   Campaign', *The Telegraph* (23 May 2015), www.telegraph.co.uk.
48 'Rainbow City, Alabama to Change Town Name to Something
   Less Faggy', Landover Baptist Church (9 September 2013),
   www.landoverbaptist.net.

SELECT BIBLIOGRAPHY

Adams, Douglas, *The Hitchhiker's Guide to the Galaxy* (London, 1979)
—, *The Original Hitchhiker Radio Scripts*, ed. Geoffrey Perkins
    (London, 1985)
Ahir, D. C., *Buddhism in Modern India* (New Delhi, 1991)
Ambros, Barbara, 'Vengeful Spirits or Loving Spiritual Companions?
    Changing Views of Animal Spirits in Contemporary Japan', *Asian
    Ethnology*, LXIX/1 (2010), pp. 35–67
Anon., *The Origin of Duty and Right in Man, Considered* (London, 1796)
Bacon, Roger, *The Opus majus of Roger Bacon*, ed. J. H. Bridges (London,
    1900), vol. II
Barlow, George, 'Awushioo: Henry Dumas at the Rainbow Sign', *Black
    American Literature Forum*, XXII/2 (1988), pp. 167–70
Barnes, R. H., 'The Rainbow in the Representations of Inhabitants of
    the Flores Area of Indonesia', *Anthropos*, LXVIII (1973), pp. 611–13
Ben-Menahem, Ari, *Historical Encyclopedia of Natural and Mathematical
    Sciences, Volume 1: Pre-science–1583 CE* (Berlin, 2009)
Blair, Walter, 'Color, Light, and Shadow in Hawthorne's Fiction',
    New England Quarterly, XV/1 (1942), pp. 74–94
Blust, Robert, 'The Origin of Dragons', *Anthropos*, XCV (2000),
    pp. 519–36
Boas, Franz, 'The Limitations of the Comparative Method of
    Anthropology', *Science*, IV/103 (1896), pp. 901–8
Boyer, C. B., 'Descartes and the Radius of the Rainbow', *Isis*, XLIII/2
    (1952), pp. 95–8
—, 'Kepler's Explanation of the Rainbow', *American Journal of Physics*,
    XVIII/6 (1950), pp. 360–66
—, 'Robert Grosseteste on the Rainbow', *Osiris*, XI (1954), pp. 247–58
Brumbaugh, R., 'The Rainbow Serpent on the Upper Sepik', *Anthropos*,
    LXXXII (1987), pp. 25–33
Buswell, R. S., Jr, and D. S. Lopez, Jr, eds, *The Princeton Dictionary
    of Buddhism* (Princeton, NJ, 2013)

Byrne, M., 'Over the Rainbow', *Fortnight*, 396 (2001), p. 8

Carter, D., *Stonewall: The Riots that Sparked the Gay Revolution* (New York, 2004)

Cheshire, F. J., 'Rainbow Magic', *Folklore*, xii/4 (1901), pp. 479–80

Child, J., 'From "Color" to "Rainbow": u.s. Strategic Planning for Latin America, 1919–1945', *Journal of Interamerican Studies and World Affairs*, xxi/2 (1979), pp. 233–59

Christiansen, R. T., 'Myth, Metaphor, and Simile', *Journal of American Folklore*, lxviii/270 (1955), pp. 417–27

Cobo, B., *History of the Inca Empire*, ed. and trans. Roland Hamilton (Austin, tx, 1979)

Crain, Jay, and Vicki Pearson-Rounds, 'A Fallen Bat, a Rainbow, and the Missing Head', *Indonesia*, 79 (2005), pp. 57–68

Crombie, A. C., *The History of Science from Augustine to Galileo* [1959] (New York, 1995)

Curcic, Slobodan, 'Divine Light: Constructing the Immaterial in Byzantine Art and Architecture', in *Architecture of the Sacred: Space, Ritual and Experience from Classical Greece to Byzantium*, ed. B. G. Wescoat and R. G. Ousterhout, (Cambridge, 2012), pp. 307–37

Dales, R. C., *The Scientific Achievement of the Middle Ages* [1973] (Philadelphia, pa, 1994)

Davis, G. L., '"Somewhere over the Rainbow . . .": Judy Garland in Neverland', *Journal of American Folklore*, cix/432 (1996), pp. 115–28

Duhem, Pierre, 'History of Physics', in *Essays in the History and Philosophy of Science*, ed. and trans. Roger Ariew and Peter Barker (Indianapolis, in, 1996)

Dunkerly, J., 'Evo Morales, the "Two Bolivias" and the Third Bolivian Revolution', *Journal of Latin American Studies*, xxxix/1 (2007), pp. 133–66

Enright, Jonathan, 'The Rainbow', *Books Ireland*, 229 (2000), pp. 75–6

Epstein, J. L., and M. L. Greenberg, 'Decomposing Newton's Rainbow', *Journal of the History of Ideas*, xlv/1 (1984), pp. 115–40

Fisher, Philip, *Wonder, the Rainbow, and the Aesthetics of Rare Experiences* (Cambridge, ma, and London, 1998)

Friesen, Abraham, 'Philipp Melanchthon (1497–1560), Wilhelm Zimmermann (1807–1878) and the Dilemma of Müntzer Historiography', *Church History*, xliii/2 (1974), pp. 164–82

Frye, N., *A Natural Perspective* (New York and London, 1965)

Gage, John, *Color and Culture: Practice and Meaning from Antiquity to Abstraction* (Berkeley, ca, and Los Angeles, ca, 1993)

Garber, M. D., 'Chymical Wonders of Light: J. Marcus Marci's Seventeenth-century Bohemian Optics', *Early Science and Medicine*, x/4 (2005), pp. 478–509

Gilman, E. B., "'All eyes": Prospero's Inverted Masque', *Renaissance Quarterly*, xxxiii/2 (1980), pp. 214–30

Goethe, J. W. von, 'Scientific Studies', in *Goethe: The Collected Works*, ed. D. Miller (Princeton, nj, 1995), vol. xii

Graziani, R., 'The "Rainbow Portrait" of Queen Elizabeth i and its Religious Symbolism', *Journal of the Warburg and Courtauld Institutes*, xxxv (1972), pp. 247–59

Habib, A., 'South Africa: The Rainbow Nation and Prospects for Consolidating Democracy', *African Journal of Political Science/Revue Africaine de Science Politique*, ii/2 (1997), pp. 15–37

Haldane, Michael, 'The Translation of the Unseen Self: Fortunatus, Mercury and the Wishing-hat', *Folklore*, cxvii/2 (2006), pp. 171–89

Harburg, Ernie, and Harold Meyerson, *Who Put the Rainbow in The Wizard of Oz?* (Ann Arbor, mi, 1993)

Hawthorne, Nathaniel, *The House of the Seven Gables* [1851] (Rockville, md, 2008)

—, *The Scarlet Letter: A Romance* (Leipzig, 1852)

—, *Twice-told Tales* (Philadelphia, pa, 1889)

Hooke, Robert, *An Attempt to Prove the Motion of the Earth from Observations* (London, 1674)

—, *Micrographia, Or Some Physiological Descriptions of Minute Bodies Made with Magnifying Glasses with Observations and Inquiries Thereupon* (London, 1665)

Hunt, J. D., *Garden and Grove: The Italian Renaissance Garden in the English Imagination, 1600–1750* (Philadelphia, pa, 1986)

Huygens, Christiaan, *Traité de la lumière* (Leiden, 1690)

Johnson, Paulette S., 'When Every Cloud is Drained of Sorrow and Anger', *Callaloo*, ii (1978), pp. 137–9

Kenneally, M., 'Autobiographical Revelation in O'Casey's "I Knock at the Door"', *Canadian Journal of Irish Studies*, vii/2 (1981), pp. 21–38

Kepler, Johannes, *Paralipomena* (Frankfurt, 1604)

Kershaw, B., 'Ecoactivist Performance: The Environment as Partner in Protest?', *Drama Review*, xlvi/1 (2002), pp. 118–30

Kidd, I. G., *Posidonius*, vol. ii: *The Commentary, Testimonia and Fragments 1–149* (Cambridge, 1988)

LaFountain, J. D., 'Colorizing New England's Burying Grounds', in A. Feeser et al., eds, *The Materiality of Color: The Production, Circulation, and Application of Dyes and Pigments, 1400–1800* (Farnham, 2012), pp. 13–28

Lande, A., 'The Decline and Fall of Quantum Dualism', *Philosophy of Science*, xxxviii/2 (1971), pp. 221–3

Landow, George P., 'Rainbows: Problematic Images of Problematic Nature', The Victorian Web, www.victorianweb.org, accessed 21 January 2016

Lang, Andrew, ed., *Yellow Fairy Book* (London, 1956)
Lang, Bernhard, 'Non-Semitic Deluge Stories and the Book of Genesis:
   A Bibliographical and Critical Survey', *Anthropos*, LXXX (1985),
   pp. 605–16
Lee, Raymond L., Jr, and Alistair B. Fraser, *The Rainbow Bridge:
   Rainbows in Art, Myth, and Science* (University Park, PA, 2001)
Lévi-Strauss, Claude, *From Honey to Ashes: Introduction to a Science of
   Mythology*, trans. John Weightman (New York, 1973), vol. II
Lindberg, D. C., 'Roger Bacon's Theory of the Rainbow: Progress or
   Regress?', *Isis*, LVII/2 (1966), pp. 235–48
Loewenstein, G., and L. T. Sanders, 'Bloc Voting, Rainbow Coalitions,
   and the Jackson Presidential Candidacy: A View from Southeast
   Texas', *Journal of Black Studies*, XVIII/1 (1987), pp. 86–96
Loewenstein, John, 'Rainbow and Serpent', *Anthropos*, LVI (1961),
   pp. 31–40
McKnight, David, *People, Countries, and the Rainbow Serpent:
   Systems of Classification among the Lardil of Mornington Island*
   (Oxford, 1999)
Mallory, J. P., and D. Q. Adams, *Oxford Introduction to Proto-Indo-
   European and the Proto-Indo-European World* (London, 2006)
Martins, A. A., 'Radio Drama for Development: ARDA and the
   "Rainbow City" Experience', *Journal of African Cultural Studies*,
   XVI/1 (2003), pp. 95–105
Mason, John, 'Yorùbá Beadwork in the Americas: Òrìṣà and Bead
   Color', *African Arts*, XXXI/1 (1998), pp. 28–35 and 94
Mead, M., 'The Mountain Arapesh II: Supernaturalism', *Anthropological
   Papers of the American Museum of Natural History*, XXXVII/3 (1940),
   pp. 317–451
Melton, J. G., and M. Laumann, eds, *Religions of the World:
   A Comprehensive Encyclopedia of Beliefs and Practices*, 2nd edn
   (Santa Barbara, CA, 2010)
Métraux, Alfred, *Myths of the Toba and Pilagá Indians of the Gran Chaco*
   (Phildaelphia, PA, 1946)
Mills, Clifford, *Where the Rainbow Ends, a Fairy Story . . . Based on the
   Fairy Play of the Same Name by Clifford Mills and John Ramsey*
   (London, n.d. [after 1910])
Milton, John, *Comus: A Maske* (London, 1637)
Moreno Vega, Marta, 'The Candomble and Eshu-Eleggua in Brazilian
   and Cuban Yoruba-based Ritual', in *Black Theatre: Ritual Perform-
   ance in the African Diaspora*, ed. P. C. Harrison et al. (Philadelphia,
   PA, 2002), pp. 153–66
Morrissey, Tracie, 'Inside the Rainbow Gulag: The Technicolor Rise and
   Fall of Lisa Frank', Jezebel, http://jezebel.com (12 December 2013),
   accessed 28 April 2017

Neruda, Pablo, *The Book of Questions*, trans. William O'Daly
    (Port Townsend, WA, 2001)
Newton, Isaac, *The Optical Papers of Isaac Newton*, ed. A. E. Shapiro
    (Cambridge, 1984), vol. I
Niman, Michael I., *People of the Rainbow: A Nomadic Utopia* (Knoxville,
    TN, 1997)
Nimetz, M., 'Shadows on the Rainbow: Machado's "Iris de la Noche"',
    Hispania, LIX/1 (1976), pp. 50–57
Norman, N. L., and K. G. Kelly, 'Landscape Politics: The Serpent Ditch
    and the Rainbow in West Africa', *American Anthropologist*, new ser.,
    CVI/1 (2004), pp. 98–110
Ockenden, R. E., 'Marco Antonio de Dominis and his Explanation of
    the Rainbow', *Isis*, XXVI/1 (1936), pp. 40–49
Ostdiek, V., and D. Bord, *Inquiry into Physics*, 7th edn (Boston, MA, 2013)
Paine, Thomas, *The Writings of Thomas Paine*, vol. III: *1791–1804*, ed. M. D.
    Conway (New York and London, 1895)
Pierce, R. H., 'The Rainbow Mosaic at Pergamon and Aristotelian
    Color Theory', *American Journal of Archaeology*, LXXII/1 (1968), p. 75
Pope, Alexander, ed., *The Works of Shakespear* (London, 1728)
Quint, H. H., 'Gaylord Wilshire and Socialism's First Congressional
    Campaign', *Pacific Historical Review*, XXVI/4 (1957), pp. 327–40
Radcliffe-Brown, A. R., 'The Rainbow-serpent Myth of Australia',
    Journal of the Royal Anthropological Institute, LVI (1926), pp. 19–25
Roberson, D., I. Davies and J. Davidoff, 'Color Categories Are Not
    Universal: Replications and New Evidence from a Stone-age
    Culture', *Journal of Experimental Psychology: General*, LXXIX/3
    (2000), pp. 369–98
Roberts, A. F., 'Chance Encounters, Ironic Collage', *African Arts*, XXV/2
    (1992), pp. 54–63 and 97–8
Roper, Alan, 'Dryden, Sunderland, and the Metamorphoses of a
    Trimmer', *Huntington Library Quarterly*, LIV/1 (1991), pp. 43–72
Saltz Merker, Gloria, 'The Rainbow Mosaic at Pergamon and
    Aristotelian Color Theory', *American Journal of Archaeology*, LXXI/1
    (1967), pp. 81–2
Sayili, Aydin M., 'The Aristotelian Explanation of the Rainbow', *Isis*,
    XXX/1 (1939), pp. 65–83
Schweizer, P. D., 'John Constable, Rainbow Science, and English Color
    Theory', *Art Bulletin*, LXIV/3 (1982), pp. 424–45
Scott, Walter, *The Poetical Works of Walter Scott, Esq.* (Edinburgh, 1820),
    vol. IV
Sherrill, C. H., 'The Five Stripes of China's Flag', *North American
    Review*, CCXI/773 (1920), pp. 517–18
Simonyi, Karoly, *A Cultural History of Physics*, trans. David Kramer
    (Boca Raton, FL, 2012)

Smith, Grafton, *The Evolution of the Dragon* (Manchester, 1919)

Spicer, E. H., *The Yaquis: A Cultural History* (Tucson, AZ, 1980)

Stuewer, R. H., 'A Critical Analysis of Newton's Work on Diffraction',
Isis, LXI/2 (1970), pp. 188–205

—, 'Was Newton's "Wave-particle Duality" Consistent with Newton's
Observations?', *Isis*, LX/3 (1969), pp. 392–4

Subramaniam, Madabusi, *At the Feet of the Master* (New Delhi, 2002)

Taylor, Douglas, 'The Dog, the Opossum and the Rainbow',
International Journal of American Linguistics, XXVII/2 (1961),
pp. 171–2

Thoreau, Henry David, *Walden and Other Writings* [1854]
(New York, 1937)

Trezise, Simon, ed., *The Cambridge Companion to Debussy* (Cambridge,
2003)

Twain, Mark, *A Tramp Abroad* [1879] (New York, 1921)

Valji, N., 'Creating the Nation: The Rise of Violent Xenophobia in the
New South Africa', unpublished master's thesis, York University
(2003)

Van Der Riet, J., 'Triumph of the Rainbow Warriors: Gender,
Nationalism and the Rugby World Cup', *Agenda: Empowering
Women for Gender Equity*, 27 (1995), pp. 98–110

van Rij, Inge, *The Other Worlds of Hector Berlioz: Travels with the
Orchestra* (Cambridge, 2015)

Venturi, Giambatista, *Commentari sopra la storia e le teorie dell'ottica*
(Bologna, 1814), vol. I

Williams, G. D., *The Cosmic Viewpoint: A Study of Seneca's 'Natural
Questions'* (Oxford, 2012)

Wood, R. W., *Physical Optics* (New York, 1928)

Young, Thomas, *A Course of Lectures on Natural Philosophy and the
Mechanical Arts* (London, 1807), vol. I

Zwerner, S. M., *The Influence of Animism on Islam: An Account of Popular
Superstitions* (New York, 1920)

ACKNOWLEDGEMENTS

Scores of people helped or encouraged me during the unexpectedly lengthy period it took me to complete this book. In particular, I would like to thank Nathan Abrams, Anna Brown, Julie Cuthbertson, Gabrielle Daniels, Peter Davidson, Gil Duran, Marc Ellington, Shana Feibel, Renske Franken-Le Clercq, Tom Freshwater, Lynne Lumsden Green, Philip Hickok, Edward Hocknell, Victoria Hodgson, Catherine Holliss, Catherine Jewkes, Kimmo Karjalainen, Heather Konkoli, Jan Leatham, Dean MacCannell, Eleanor MacCannell, Jason MacCannell, Juliet MacCannell, Hesse McGraw, Michael Jones McKean, Lex Meddings, Gilbert Ramsay, Paul Rushton, Hugh Salvesen, Francesca Sanchez, Philip Schwyzer, Paul Shanks, Jolyon Spencer and Jessica Wheelis. Any errors, however, are my responsibility alone.

PHOTO ACKNOWLEDGEMENTS

The author and the publishers wish to express their thanks to the below sources of illustrative material and /or permission to reproduce it.

Alamy: p. 113 (ART Collection); © 2008 by Baccara, CC BY-SA 3.0: p. 21; © 2009 by Salvatore Barbera, CC BY 2.0: p. 151; © 2012 by Harvey Barrison, CC BY-SA 2.0: pp. 92–3; The British Library, London: p. 37; © Michael Carian, CC BY-SA 2.0: p. 156; © Christopher Cox, CIRES/University of Colorado, CC BY-SA 2.0: p. 11; © DACS 2017: pp. 134–5; © 2013 by Asher Floyd, CC BY-SA 3.0: p. 104; © 2015 by François Goglins, CC BY-SA 4.0: p. 74; © Sorcha A. Hazelton, CC BY-SA 4.0: p. 154; © 2009 by Heiser, CC BY-SA 3.0: p. 23; © 2005 by Brocken Inaglory, CC BY-SA 3.0: p. 20; © 2012 by Adam Jones, CC BY-SA 2.0: p. 35; © www.rodjonesphotography.co.uk, CC BY 2.0: p. 17; Courtesy of and © 2014 by Jan Leatham: p. 15; © 2015 by Kalki, CC BY-SA 4.0: p. 79; Library of Congress, Prints and Photographs Division, Washington, DC: pp. 6 (The Jon B. Lovelace Collection of California Photographs in Carol M. Highsmith's America Project), 16, 27, 28, 36, 39, 42, 47, 48, 50, 54, 68, 71, 75, 81, 85, 92, 106, 119, 122, 123, 124 (The Jon B. Lovelace Collection of California Photographs in Carol M. Highsmith's America Project), 129, 140, 141, 142, 143, 144, 145, 147, 148, 159; © Kiwi Flickr, CC BY 2.0: p. 126; © 2011 by Robert Luna, CC BY 3.0: p. 139; © 2007 by Ralf Lotys, CC BY 3.0: p. 136; © 2015 by Daniel MacCannell: p. 157; Courtesy of and © Michael Jones McKean: p. 125; © Attila Magyar, CC BY-SA 2.0: p. 24; © Luis Marina, CC BY 2.0: p. 32; © 2010 by Mike Peel, www.mikepeel.net, CC BY-SA 4.0: p. 78; © 2016 by José Luiz Bernardes Ribeiro, CC BY-SA 4.0: p. 107; Courtesy of and © Paul Rushton, www.paulrushtonphotography.com: pp. 9, 14, 18–19, 22; Shutterstock: pp. 58 (Vectomart), 103 (Enriscapes); © 2005 by snowyowls, CC BY-SA 2.0: p. 63; © Bjørn Christian Tørrissen, CC BY-SA 3.0: p. 146; Trapped in Suburbia: p. 150; © Trogain, CC BY-SA 3.0: p. 57; © Nusha Uporabnik, CC BY-SA 3.0: p. 12; Walters Museum of Art, Baltimore: p. 108; The Wellcome Library, London: pp. 26, 51 (both items CC BY 4.0);

INDEX